Running Android

Using Your Phone and Tablet for Work and Play

Bruce Grubbs

Bright Angel Press

Flagstaff, Arizona

Running Android

Using Your Phone and Tablet for Work and Play

© Copyright 2015-2016 Bruce Grubbs

Updated May 2016 for Android Marshmallow 6.0.1

Bright Angel Press

Flagstaff, Arizona

www.BrightAngelPress.com

ISBN: 978-0-9899298-2-0

Contents

Acknowledgments

I would like to thank all the Android users and experts out there who contribute to the various Android forums and share their insights with us all. Big thanks to Peter Levine for an excellent job of copy editing as always. And thanks to Duart Martin for offering valuable suggestions and for her untiring support of my writing projects.

Introduction

Congratulations! If you're reading this book, you either own or are seriously thinking of buying an Android-based tablet or phone. Google's wildly popular Android operating system dominates the portable device market- it is now installed on over 80% of tablets and phones worldwide, and also has by far the most third party apps (applications). If you'd like to have a phone or a tablet that you can customize to work your way and not be locked into a single company's idea of how things should work, Android is for you.

I've updated this book to cover the current version of Android, 6.0.1, code named Marshmallow. All of the instructions, descriptions, examples, and screenshots have been taken from my phone and my tablet running the latest release available for each device.

This book is based on my earlier book, *Google Nexus 2013: Making Your Android Tablet Work for You*, which covers Android Kitkat. Since later versions of Android are often slow to roll out across devices, I've decided not to revise the Nexus book, and instead keep it available as a reference for KitKat users.

The purpose of this book is to help you get the most out of your Android phone and tablet by showing you how to do practical, everyday things. Although the book has a getting started chapter to help you get up and running with your new device right out of the box, it does not try to replicate the instructions that come with every device.

Google, the developer of the Android operating system, wisely places few limitations on how the user interface may be customized, and so many phones and tablets run versions of Android that are highly customized. However, the underlying Android is still the same and all these devices work basically the same way. There's a pretty good chance that if an app works on your phone, it will also work on your tablet, so you can work or play the same way on both devices.

Obviously, some apps just can't work on both phones and tablets. Examples are dialers and text messaging apps, which are useless on a tablet. A contact manager, on the other hand, can be installed on both so that if you want to look up a friend and send them an email, you can do it the same way on both your phone and tablet.

Using This Book

To help you make the best use of your device, I refer to specific apps and websites throughout the book. So that you can easily find these resources, the book has direct links within the text. In the ebook version, these are hyperlinks that take you directly to the website I'm referencing. All hyperlinks in the ebook are in blue (for those reading on color devices) and underlined, following the standard web convention. An example is www.GrandCanyonGuide.net.

The print version has printed links that you can type into your web browser. Long link addresses are shortened with bit.ly to make them easier to type in. For example, http://amzn.to/1bkPwxO takes you to an Amazon product detail page that has a very long link address. After you follow a link from your device, you can return to the previous page by tapping the Back icon at the bottom of the screen.

Following hyperlinks from a phone or tablet requires that it has a wireless Internet connection, either Wi-Fi or LTE (3G or 4G cellular data), in order to display the target Web page.

Since both versions of the book refer you to other chapters and sections within the book, the ebook has internal hyperlinks so you can jump directly to the reference. To return to your previous place, just hit the Back button or icon on your e-reader or tablet. For example, if you click on Getting Started, you'll jump to the "Getting Started" chapter. The print book refers you to chapter and section names.

What to Buy?

Since this book isn't a buyer's guide, and there are so many Android phones and tablets on the market now, I'll just touch on a few considerations to think about when shopping.

How Pure is Your 'Droid?

As mentioned above, most manufacturers modify the user interface on their Android devices. Google is the exception, since they want to showcase their Android system. Their Nexus line of tablets and their smart phones have the purest and latest Android. But there are many reasons to go with a phone or tablet from another company.

Cell Coverage

In the case of a phone, as well as a tablet with LTE (cellular) data capability, voice and data coverage are usually a prime consideration. If you're primarily an urban dweller, then you're in luck. Most cell phone companies provide solid coverage of major urban areas. On the other hand, if you travel a lot, especially to remote areas, then compare coverage maps carefully. In Arizona, where I live, there are large areas with no coverage at all, and one company clearly stands out as having the best rural coverage. As a backcountry hiker and as a charter pilot flying to remote airstrips, I really only have one choice of cell company.

Memory and Storage Size

You never have too much memory on your phone or tablet, especially if you listen to music or watch videos, so always buy the model with the most built-in storage- unless you can find a phone or tablet with a micro-SD card slot. This slot lets you add external memory cards, theoretically up to 2tb on the latest devices. In practice, the largest cards available hold 200gb, but this is still a massive expansion for a device with 32gb or less of built-in memory.

For several years, the trend has been against including micro-SD card slots, but now that Android Marshmallow fully supports external memory for apps and data, it's likely that more phones and tablets will include micro-SD slots.

Wi-Fi or LTE on Your Tablet?

Since public, free Wi-Fi Internet is so prevalent, Wi-Fi tablets are probably the best choice for most users. Also, don't forget that you can use a smart phone to provide a Wi-Fi hotspot for your tablet anywhere your phone has an LTE data connection.

The main reasons to get an LTE tablet would be if your home doesn't have Wi-Fi or you frequently travel in rural areas that don't have much in the way of public Wi-Fi. Remember that you will need a data plan with a cellular phone company to connect to the Internet via LTE. If you already have a smart phone, it's usually much cheaper to share a data plan with your phone than to buy a separate plan.

Be careful with apps that use large amounts of data, such as movies, when connected via LTE. You can quickly go through your monthly data allowance and then pay high rates for the excess. Most cell phone companies allow you to set up alerts so you'll get an email or text when you approach your data limits.

Where to Buy?

You can buy Android phones and tablets from a cellular company, or from online and physical retailers. There are several major advantages to buying from a cell provider. First, the sales rep will set up your device for you and you'll walk out of the store with your new phone or tablet ready to use. Also, most cell companies deeply discount devices that you buy through them, as long as you're willing to sign a one or two year contract. Finally, most cell companies offer free or low cost upgrades after a year, so you can get a new phone or the latest model.

If you buy your phone or tablet from a retailer, you'll pay retail price and you'll have to set it up yourself. The advantages are that you're free to buy any Android phone or tablet on the market and aren't limited to the models that your cell company carries. Phones and tablets with LTE (cell data) that aren't programmed for a specific cell company are known as unlocked devices. You then take your unlocked device to the cell company, where they provide a SIM card. The SIM card programs the phone or tablet for that cell network.

If you're buying a Wi-Fi-only tablet, there's no real advantage to buying from a cell company (unless they do a bundled discount with a phone), since the tablet won't be on a cell network anyway. And as mentioned earlier, you can always use your phone as a Wi-Fi hotspot to give your tablet (and any other Wi-Fi device, such as a laptop) Internet access, as long as your phone is on the cell data network.

Getting Started

The first thing you should do with your shiny new Android phone or tablet is charge it. New devices come with the battery about half-charged (this prolongs battery life in storage), and when you first turn it on, it will almost certainly download and install a system update. While it's unlikely, it is possible that the battery will fail part way through the update, turning your new device into a brick.

Instead, unpack the charger and plug it into a wall outlet, then plug the cable into the charger and into the micro USB port on the bottom edge of the tablet. Both plugs fit only one way. Keep your new phone or tablet on the charger for a few hours.

This device is charging, as shown by the lightning bolt symbol on the battery icon on the Status bar

Turning your Phone or Tablet On for the First Time

If you bought an unlocked LTE tablet or an unlocked phone, you'll need to take your device to a cell phone company to buy cell service and get a SIM card. Normally, the sales rep sets up your phone or tablet for you. In some cases, when you buy a device with a cell contract from an online retailer, you'll have to insert the SIM card and do the setup yourself. In this case, follow the instructions that came with the phone or tablet.

Once your tablet or phone is fully charged, or at least connected to the charger, you can turn it on by pressing and holding the power button until the screen wakes up. If your device was set up by a cell company, you should be ready to start using it. If not, follow the on-screen set-up instructions. When set-up is complete, you'll be on the Home screen.

If you have a Wi-Fi-only tablet, you'll need a Wi-Fi connection to complete the setup.

Getting Around on Android

Physical Buttons and Ports

Most Android devices have three physical buttons and two ports. The power button is used to turn on the device when it has been completely powered down, as mentioned above. When the device is on but sleeping (the normal state, with a blank screen), a momentary press of the power button wakes the screen from sleep mode. When the screen is on, pressing and holding the power button brings up the power pop-up, which gives you the option of powering off the device completely.

The physical volume control buttons work even when the screen is sleeping, so you can adjust the playback volume of music without waking up the screen. When the screen is awake, a volume control slider pops up on the screen. By default, the slide controls the alarm or ringer volume, but if you're listing to music or a video, the default controls the listening volume. You can select other volume settings by touching the drop-down at the right.

As mentioned above, the micro-USB port is used for charging the device. It can also be used to connect the phone or tablet to a computer to transfer files, such as music and photos.

Some Android tablets and phones also have a slot for an SD or Micro-SD card, which allows you to expand the memory so that you store more photos, videos, and music on the device. Phones and LTE tablets also have a slot for a SIM card, which programs the device for use with a specific cellular carrier.

Home Screen

As the name implies, the Home screen is your starting point in Android. You can actually have more than one Home screen- to see them, swipe left or right. But you can always return to the main Home screen by tapping the Home icon at the bottom center of the screen.

Touch Screen Navigation

Except for the three physical buttons, you use Android phones and tablets by tapping, swiping, and gesturing on the touch screen. To activate something, touch it. To enter text, touch where you want to type, which activates the on-screen keyboard. For example, when you touch a search icon, the search text field appears and the keyboard pops up. When you're finished typing, tap the Search icon on the keyboard, or Go, and the keyboard disappears. You can also dismiss the keyboard by tapping the Back icon.

You can select an item by touching and holding it. Options will pop up. On the keyboard, touching and holding selects alternate characters. Touch, hold, and drag is used to move things around. An example is copying an app from the Apps Drawer (see below) to the Home screen. When you lift your finger, the app icon appears there.

Touch the screen and then slide your finger up, down or sideways without pausing to move between home screens or to pan a map. Touch with two fingers at once, and then spread your fingers to zoom out and pinch your fingers to zoom in. Rotate your two fingers to rotate the screen. Google Maps is an excellent app to practice these skills. Not all gestures work in all apps.

Screen Orientation

Android devices automatically switch from portrait to landscape mode when you rotate the device (unless you've locked the screen in Settings.) To avoid confusion, all of the screen descriptions in this book assume that the screen is in portrait orientation, unless noted.

Lock Screen

Android devices go to sleep and the screen goes blank after a short period of inactivity, which greatly extends battery life. The device goes to sleep even if it's on a charger, because it charges faster with the screen off.

Most Android devices default to a slide lock screen. To turn it on after the tablet has gone to sleep, press the power button momentarily, then swipe the Lock icon upward. There are other lock screens you can set- see Securing Your Device.

Navigation Bar

At the bottom edge of the screen, you'll see the Navigation bar, which always contains three icons. If they aren't visible, which can occur with some full-screen apps such as games and reading apps, touch the center of the screen.

The center icon is the Home icon and tapping it always takes you to the last Home screen you were on. If you have more than one Home screen, tapping Home a second time always takes you to the main Home screen.

The Back button on the left takes you back to the previous screen that you were using. Tapping it repeatedly takes you back further.

The Recent button is the one on the right, and touching it shows you a scrolling stack of the most recent screens you used. You can scroll rapidly through the stack to find a recent app, without having to open the app again from the Home screen or App Tray.

The Recent and Back buttons are your friends. Whether the last place you visited was a web page, your email, a map, or some other app, the Back icon can take you right back there. And the Recent button save a lot of time when you're repeatedly flipping between apps- say from Gmail to the web, or Wikipedia to a dictionary.

Even though an Android device can't show more than one app at a time on its screen, many apps continue to run in the background, and if you open them from the Recent button you'll be right where you left off.

Favorites Tray

The Favorites tray is at the bottom of every Home screen, just above the Navigation bar, and contains apps that you use all the time. The Apps Drawer icon, a small circle filled with dots, is always located in the center of the Favorites tray, and tapping it takes you a screen of icons for all the apps and widgets installed on your device, sorted alphabetically.

The Home screen and Favorites tray can be customized with the apps and widgets you prefer. In addition, you can customize the Lock screen, the screen you see when you power up. For details, see Customizing Your Device.

Tap the Apps Drawer icon now and you'll get a screen full of app icons. Actually, there are several screens of apps. The App Drawer always contains all of the apps installed on your device and is the place to go to when you can't find an app anywhere else.

The Home screen on a phone, with the Favorites Tray at the bottom. The Apps Drawer is the circle with dots in the middle of the Favorites Tray. This Home screen has been customized with Nova Launcher

Notification Area

The Notification area at the left top edge of the screen is where small icons appear to notify you of app updates, email, calendar events, and other notifications. Swipe down from the top of the screen to view details of each notification. Tap a notification to open its app. For example, tapping on an email notification opens your Gmail or email app. Once you open a notification, the notification icon disappears from the Notifications area. You can tap the icon at the end of the list to dismiss all of the notifications. Or tap anywhere else to hide the list without dismissing the notification icons.

Status Area

The Status area at the upper right corner shows priority notification, Bluetooth, Wi-Fi, and battery status, as well as the time. Swipe down from the top of the screen with two fingers (or twice with one finger) to open the Quick Settings panel, and tap anywhere else to hide it. From Quick Settings, you can adjust screen brightness, turn Wi-Fi, Bluetooth, and Airplane Mode on or off, lock the screen orientation, and toggle location services. There may be more options, depending on your device.

Making Phone Calls

If you have an Android phone, the left-most icon in the Favorites Tray is usually a Phone icon. Tap it to bring up a dialer keypad. The pre-installed Phone app also lets you get a list of recent outgoing and incoming calls, as well as a list of favorites you've set in the Contacts app.

Texting

A Messaging icon on the Favorites Tray lets you send and receive text messages. Often, an easier way to make phone calls or text is from the Contacts app. See the chapter Managing Contacts.

Copy, Cut, and Paste

Most apps allow you to select text, modify the selection, and then cut, copy, or paste it. To select, long-press and drag over the approximate text. Then drag the handles to modify your selection. If you drag to expand the selection, it expands by whole words. If you drag to narrow the selection, it shrinks one character at at time. Then use the pop-up to cut or copy your selection. To paste, long-press in the text, then drag the handle to the exact place you wish to insert the copied text. Then press the paste icon in the pop-up.

Using the Flashlight

You can use the camera flash LED found on most devices as a flashlight by tapping the Flashlight app in the app drawer, or from any screen (including the lock screen) by making a chopping motion with the phone twice. Chop twice again to turn it off.

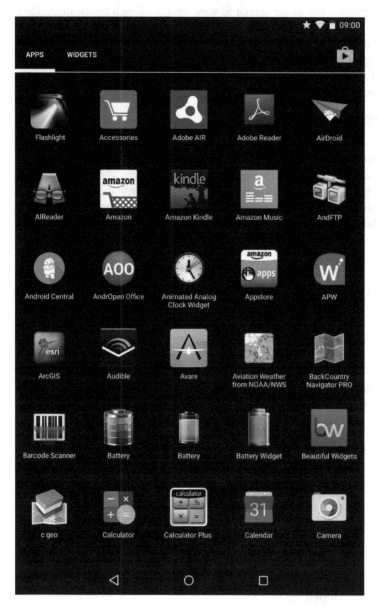

The open Apps Drawer, with app icons sorted alphabetically. To see the rest of your apps, swipe left. Widgets are shown after apps. The Navigation bar is at the bottom of the screen

Extending with Apps

Apps and Widgets

All Android devices come with a number of pre-installed apps (short for "applications") and widgets that let you do many things with your phone or tablet, such as browsing the Internet with Chrome, checking email with the Gmail and Email apps, staying active on social networks with the Facebook and Twitter apps, reading books and magazines with Play Books, Kindle for Android and Play Magazines, watching movies and TV shows with Play Movies and TV, listening to music with Play Music, listening to books with Audible Audiobooks, finding places and getting directions with Google Maps, managing appointments with Calendar and contacts with Contacts+, and many more.

In addition, the Google Play Store offers tens of thousands of apps that offer alternatives to the pre-installed apps, as well as adding many new capabilities to Android. The following chapters describe many of the pre-installed apps as well as some of the best apps from Google Play while showing you how to use your phone or tablet to do things.

You can shop the Google Play Store from the Play Store app that is pre-installed on the device, or from Google Play on your computer (https://play.google.com/store/apps).

While we're at it, what's the difference between an app and a widget? An app is a computer program that runs on your Android device. Apps are represented by icons on the screen of the tablet. Apps are usually started by tapping the icon. Long-pressing the icon brings up a pop-up menu of options that let you edit settings, remove the icon from its current location, get info on the app, and uninstall it.

A widget is a graphic that may be as small as an icon or fill the entire screen. Most widgets are interactive and present more information than can be displayed in an icon. Some apps come with their own widgets, and you can also buy sets of widgets.

Good examples are the weather widgets described in Keeping up with the Weather. Tapping a widget usually opens the parent app but some apps can be set to open other apps. Tapping the clock and date widgets that are installed on the Home screen by default open the alarm app and Google Calendar app.

Multiple copies of app icons and widgets can be installed on different screens, allowing to you customize the tablet to your needs and preferences. The "original" app icons and widgets are always accessible from the Apps Drawer icon at the bottom of the Home screen- it's the circle. Tapping this icon opens a screen with all your apps sorted alphabetically. Since even a new Android device has more apps than can fit on one screen, swipe left to see more apps.

After the apps, you'll see the widget screens. You can jump directly to the app or widget screens by tapping Apps or Widgets at the top left corner of the screen.

Both apps and widgets are installed on your Home screen by long-pressing and dragging them to the Home screen or Favorites tray at the bottom of the screen. If you have more than one home screen or Favorites tray, drag left or right to move to another screen. Dropping a widget usually opens a configuration or settings pop-up.

Apps can be opened directly from the Apps Drawer, the Home screens, or the Favorites tray, but widgets can't be opened until they are installed on your Home screen.

Free vs. Paid Apps

Apps are either free or paid. In many cases, the free app is intended to let you try the key features of the app to be sure it works on your device and does what you want before you spend any money. Such "trial" apps often lack some key features that are available in the paid version of the app.

Other free apps differ from the paid version only in that they are ad-supported. This usually means that ads will appear along the top or bottom of the screen or in pop-ups in a corner.

Some apps are available only in a paid version, but most are inexpensive compared to commercial desktop programs. And Google Play provides a refund button right after you make your purchase so you can cancel if you buy an app by mistake.

Games

Games also come in free and paid versions. Free games often are limited versions of the full game, or else require you to start buying things to progress beyond a certain level. Other games are totally free but are ad-supported. Games are apps, by the way, but are so popular that the Play Store has a separate place for them.

Finding Apps

When you're not sure what you want, or if you just want to see what's new or hot, you can browse the App Store on Google Play. The home screen of the App Store lists categories such as "Play Picks", "Recommended for You", "Apps to Watch", "Like Recent Installs", "App Highlights", "Get Things Done", and several more.

You can also browse apps by categories such as business, comics, communication, and many more by selecting "Categories". In addition, you can look at the best selling apps by selecting "Top Charts" and the latest apps with "New Releases."

Reviews

Reviews of apps on Google Play are an excellent way to decide whether an app will do what you need. Don't let the occasional one or two star review put you off.

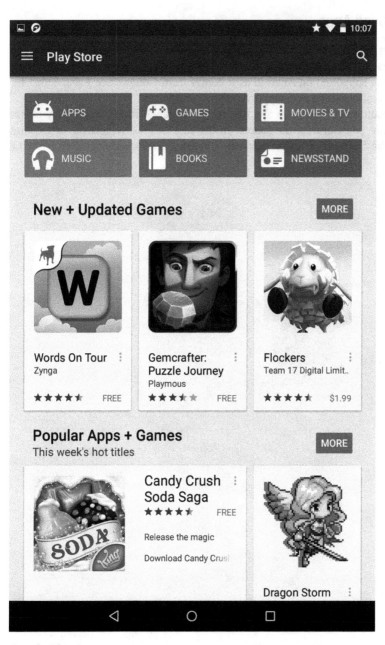

Google Play Store

Running Android

There's always someone who's unhappy, often for reasons unrelated to the purpose of the app. But if an app gets a lot of reviews complaining about the same problem, then you should pay attention.

Unapproved Apps

Sometimes it's necessary to install apps from sources other than Google Play. An example is installing the Adobe Flash app so that you can watch Flash video, which is not supported on Android as it comes from Google. See Flash Video for an example of installing an unapproved app.

You should never install an app that is not available from Google Play unless you are sure you can trust the source. In the case of Flash video, the app comes from Adobe Systems, a major software company, so it's reasonable to trust the app.

Lock screen. Swipe up on the padlock to unlock, swipe right on the microphone to start a voice search, and swipe left on the camera icon to start the camera app

Securing Your Device

Right out of the box, your Android phone or tablet has access to your Google account. Or it should, if you want to take advantage of the integration of Google apps such as Search, Gmail, Calendar, YouTube, Drive, News, and others.

If you use your Android device to connect with social networks such as Facebook and Twitter, the device will have access to those accounts. And if you let Chrome and other browsers remember passwords, then your device will have direct access to websites such as Amazon.

For all these reasons, you need to restrict access to your Android phone or tablet. There are two basic approaches. The first is to not let any Android app remember passwords. This is the most secure, but is also a total pain. You'll spend more time entering passwords than doing anything useful.

The second method is to turn on the lock screen on the phone or tablet. With the lock screen secured with a password, no one can use your device without doing a factory reset- and that wipes out all installed applications and their data, including saved logins and passwords.

The best approach is a combination of the two. Don't let critical apps and websites such as banking apps remember your user ID or password.

Some websites handle security very well. A shining example is Amazon. Although it keeps you logged in so you can see your browsing history, put items in your wish lists, and save items in your shopping cart, Amazon asks you to log in again whenever you do something critical, such as check on past orders or place a new order.

Sleep Mode

Don't confuse screen locking with sleep mode. The screen automatically turns off after a set time passes without any screen touches or tablet motion. The default is one minute but you can change this in Settings, Device, Display.

If you haven't enabled screen locking, pressing the power button wakes up the device on the last screen you were using.

Screen Lock

Screen locking options are found under Settings, Personal, Security, Screen Security. There are five options for locking the screen.

None

This option leaves the device totally open. As described above, the device returns to the screen you were on as soon as you press the power button to wake it from sleep mode. Only use this option in a secure environment where no one else has access to your tablet, such as a work place.

Slide

With this option set, the device wakes to a padlock icon, which you have to swipe to the right to return to the last open screen. Slide only prevents accidental screen touches and doesn't add any security.

Pattern

With Pattern unlock, the lock screen shows a grid of nine dots, and you draw a pattern through the dots to unlock the screen. During setup, you draw the pattern twice to confirm. For the best security, make the pattern complex and uncheck the option to make the pattern visible. If the pattern is visible, someone looking at your screen can see the pattern.

You might want to make the pattern visible at first while you learn how to draw the recorded pattern. The pattern is red if drawn incorrectly, and green when correct.

PIN

This option lets you set a numeric Personal Identification Number. To unlock the screen, you have to enter the number correctly on the PIN pad and tap Enter. Never use an easily guessed number such as your house number, birth date, or phone number. The best PIN is a number that means something to you but that no one else will associate with you.

Password

This is the most secure option, because you can enter a long password containing a mix of upper and lower case letters, numbers, and special characters. For the best security, change your password often and don't use a password that is easily guessed. Names of pets, children, spouses, your home town or street are easily guessed. As with a PIN, the best password is a complicated word or phrase that means something to you but that others will not associate with you.

Other Security Options

The Security settings screen has a number of options you can set to enhance the safety of the personal information on your tablet or phone.

Automatically Lock

If you have Pattern, PIN, or Password screen lock set, you can choose the interval after the screen goes to sleep before the screen locks. The default is five seconds. Set the interval to the shortest period that lets you wake up the screen again without having to unlock it while you are using your device.

Power Button Instantly Locks

This is especially useful if you have set a long Automatic Lock interval. If you need to leave your device unattended momentarily in a place where others might have access to it, you can press the power button to lock it instantly. This is also useful before putting the device in a case or bag where accidental screen touches are possible.

This feature also works well with tablet covers which automatically turn the tablet off when the cover is closed, and on when they are opened.

Lock Screen Message

It's a good idea to enter your name and contact information, such as your cell phone number and/or email address, into the Lock Screen Message. This information is shown on the lock screen and increases the chance that a lost device will be returned.

Smart Lock

New in Marshmallow, Smart Lock keeps your phone or tablet unlocked under trusted conditions, such as when when it's connected to a Bluetooth headset or NFC device. You can also keep the device unlocked at trusted locations such as home and work. Another option is to keep it unlocked when a trusted face is near, but face recognition could be fooled so this is not as secure. Likewise, you can unlock your device by voice. Finally, on-body detection keeps your device unlocked when it is close to your body, such as in your hand, your pocket, or handbag. Not all devices support this feature.

Encrypt Device

For the ultimate security, you can use this option to encrypt the contents of the device. To use encryption, you must set a PIN or password. Encryption is irreversible- the only way to remove it is to do a factory reset, which erases all personal data and removes all installed apps. Most users won't need this level of security unless their organization requires it.

To encrypt your device, set a PIN or password, charge it fully and leave it plugged into the charger. Then go to Settings, Personal, Security, Encrypt Tablet. You can't encrypt your device unless it is plugged into the charger. Encryption takes an hour or more, and if the device loses power during the process, some of your data could be lost.

Make Passwords Visible

Since typing passwords on a screen keyboard can take some getting used to, you have the option of making passwords visible as you type them. Once you get used to the screen keyboard, uncheck this option for the best security.

Device Administrators

This setting allows you to control which apps can act as device administrators. Use this carefully as it gives apps full control over unlocking passwords and allows apps to perform a factory reset and erase all your data and apps.

Android Device Manager

This is a built in app. By default, it can be used to locate your device if it is lost or stolen. To use this feature, location access must be turned on under Settings, Personal, Location Access. The location will be more accurate if GPS Satellites are turned on.

To locate your device, go to www.google.com/android/devicemanager. The location of your device will be shown on Google Maps, along with the accuracy of the location and date it was registered.

If you mislaid your device in your home or office, you can ring it to help you find it. Just click Ring on the Android Device Manager page. The device will ring for five minutes or until you press the power button.

You can optionally allow Android Device Manager to lock or erase your device remotely. To do this, activate Android Device Manager under Settings, Personal, Security, Device Administrators, Android Device Manager.

AirDroid

AirDroid is an example of an app that can act as a device administrator if you activate it. AirDroid is an app that lets you wirelessly transfer files and administer your device from a web page on your computer. For details, see Airdroid.

Unknown Sources

Checking this option allows you to install apps from sources other than Google Play. An example is installing the Adobe Flash app to enable web browsers such as Dolphin to play Flash Video. See Flash Video.

Normally, you should leave Unknown Sources unchecked.

Verify Apps

Normally, the Android operating system scans new apps and warns you if they might harm the devicet. You should leave this checked.

Credential Storage

These options are mainly used in a large organization using systems such as Virtual Private Networks. Your network administrator normally provides credentials if they are needed on your device.

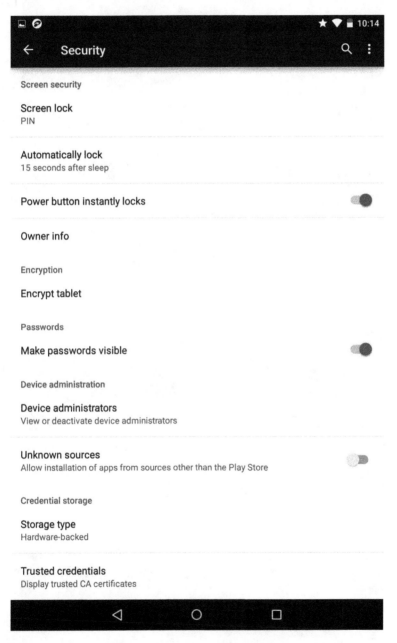

Security settings screen

Connecting to the Internet

Many things you do on your phone or tablet require an Internet connection. If you have access to Wi-Fi, use it, as it's generally faster than LTE and avoids using up your data allowance.

Airplane Mode

On the other hand, you can greatly extend battery life by turning wireless (Wi-Fi and LTE) off when you don't need an Internet connection. This is especially effective in rural and fringe areas where your phone or tablet automatically boosts its wireless power while trying to stay connected. The easiest way to turn wireless off is to swipe down from the top of the screen with two fingers to show Quick Settings, then turn Airplane Mode on by tapping its icon. This turns off all the power-hungry transmitters in your device, including Bluetooth. You should also turn on Airplane Mode on aircraft when directed by the flight crew.

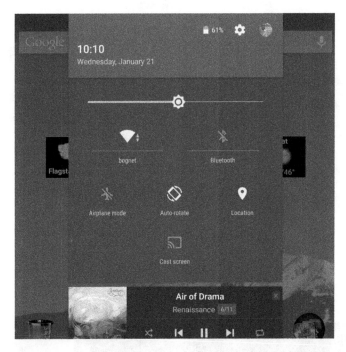

The Quick Settings pull-down, which gives you easy access to settings used most often, such as Airplane mode

Wi-Fi setup screen. The tablet is connected to Bognet

Wi-Fi Settings

To access Wi-Fi settings, swipe down from the top of the screen with two fingers to open Quick Settings, then tap the Wi-Fi icon. The Wi-Fi screen shows a list of Wi-Fi networks. The network being used, if any, is at the top of the list, followed by networks that are in range in order of signal strength. Below that, Wi-Fi networks that the tablet has used in the past are listed alphabetically.

To connect to any network that is in range, tap it. You'll be prompted for a password if the network is secured.

If you have a choice of networks, always use the one with the strongest signal. Data transfers faster when the signal is strong and you'll have fewer problems with dropped connections.

Like a laptop computer, Android remembers all the Wi-Fi hotspots it's ever used (unless you tell it to forget a hotspot) and reconnects automatically when you're in range.

Finding Wi-Fi Hotspots

There are several apps that make it easier to find Wi-Fi connections. One of the best is Wi-Fi Finder (http://bit.ly/18fwfw6), which not only shows you networks within range but also displays a map or a list of free and paid public Wi-Fi near your present location. To use the public Wi-Fi finder most effectively, the app allows you to download a worldwide database so you can find Wi-Fi hotspots when you're not connected to Wi-Fi. After all, that's when you'll be looking for a hotspot! In list mode, Wi-Fi Finder also shows whether or not a hotspot is free.

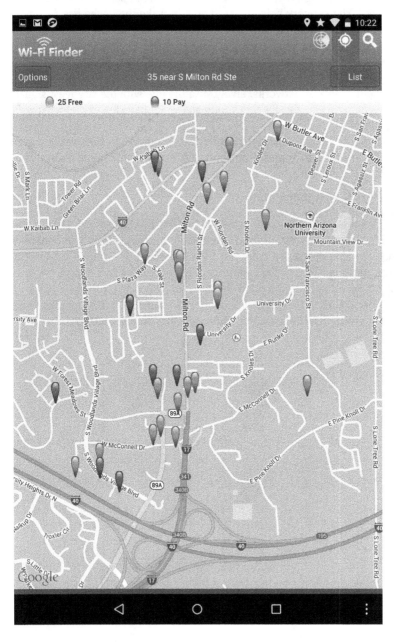

Wi-Fi Finder with local free and paid hotspots

Other Wi-Fi Apps

Speedtest (http://bit.ly/1439Hxt) lets you test the speed of your Internet connection. This is especially useful if you have a choice of more than one Wi-Fi network, or you suspect there are problems with the network you're using.

Wi-Fi Analyzer (http://bit.ly/1aIftMv) lets you look at the Wi-Fi networks around you in several useful ways.

You can display a channel graph, which shows all the networks in range by channel and signal strength. Wi-Fi automatically chooses between 14 different radio channels and usually picks the least busy channel. However, if signals are weak and there are many Wi-Fi networks in range, two or more networks may occupy the same channel. The channel graph clearly shows this and helps you choose a network with the strongest signal and least interference.

A time graph shows Wi-Fi network signal strength over time. This will be useful if you're in a dense urban area with many Wi-Fi hotspots and you're trying to find the best one. It can also alert you to problems with a Wi-Fi hotspot you're already using.

The channel rating screen is especially useful in finding the best channel for your home Wi-Fi network, or any network where you have control of the wireless router. From this screen, tap the Menu icon and select your network. The channel ratings are now shown for all channels. If you want, you can change the channel used by your Wi-Fi access point to the best channel shown. The method for doing this depends on your specific Wi-Fi access point- check its manual for instructions. Also, you should check the recommended channel several times during the day, especially in residential areas. Some people turn their Wi-Fi access points off when they're not home, so the best channel can vary.

The AP list shows a list of Wi-Fi hotspots in graphic form with signal strength bars. You can tap on a hotspot to connect directly to it. This requires another app by the same developer, Wi-Fi Connector (http://bit.ly/18s6AkA). If it's not installed, you'll be asked if you want to install it.

The last view, signal meter, shows an analog meter view of the network that your phone or tablet is currently using.

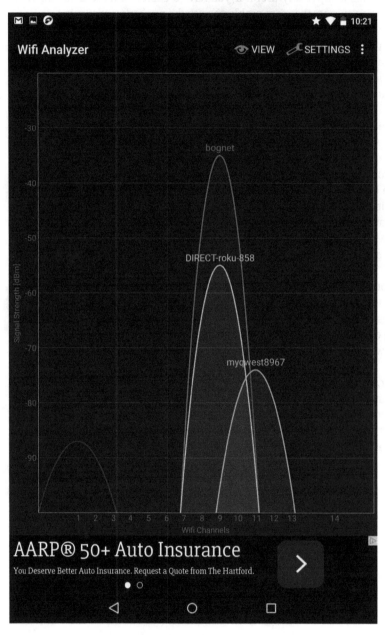

Wi-Fi Analyzer channel graph screen

Finding Things

Your phone or tablet comes with a Google Search widget at the top of the Home screen. Combined with a feature called Google Now, this is a powerful way to find places, get directions, look up websites and get information, and keep track of events from your Google calendar.

Google Search

You can search by tapping the search bar to bring up the on-screen keyboard or by tapping the microphone icon at the right end of the search bar to do a voice search. If you tap the search bar, you can then say "OK Google" to start a voice search. You can also enable voice searches from within most apps, even when the Google Search widget is hidden.

To set this up, open Quick Settings, tap the Settings icon, and open Language & Input. Under Speech, tap Voice input. Tap to enable Enhanced Google services, then tap the setup icon at the right. Now, tap From any screen to enable it. Finally, you'll be asked to train the device by saying "OK Google" three times. Tap Finish when you're done, and then Home to return to the Home screen.

You can also start a voice search from the lock screen by swiping right on the microphone icon at the lower left corner.

If you do a voice search, the device speaks the results. The spoken results vary depending on the nature of the search. If Google Search doesn't understand your request, it will present you with a list of results. Just tap the result you want, or try a new search.

For example, typing or saying "hotels" brings up a page of Google results, including a map and a list of hotels near your present location. If you want to search for hotels in a different location, just add the location. Searching for "hotels near Grand Canyon" displays a list of hotels near Grand Canyon.

You can be even more specific and provide a street address for the search location. In general, the more information you provide, the better the search results.

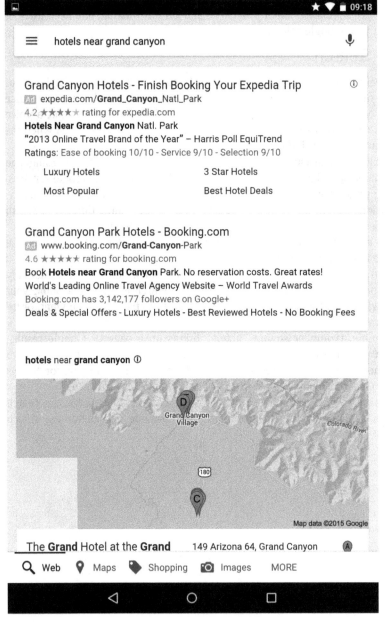

Google Search results for hotels near Grand Canyon

Getting Directions

Just type or speak the name or address of your destination. For example. "get directions to Phoenix Sky Harbor Airport" results in a map of the route and a list of directions. Just tap the "Start" icon to start navigating toward your destination. The device switches to a 3D map view and starts giving voice and on-screen turn by turn directions.

For more on navigation, see Finding Your Way With Maps and GPS.

More Searching

Google Search is incredibly flexible and can be used to search for information on nearly anything, just like Google on a computer. Here are some examples:

- What is the current weather at JFK airport?

- What time does British Airways depart London for Los Angeles?

- What time is it in Fairbanks, Alaska?

- When is moon rise?

- Where's the nearest supermarket?

- What is 3.5 times 4.6?

- Convert 100 US dollars to Canadian dollars

- How do you say "hotel" in French?

- When is the next Winter Olympics?

- Show me pictures of Moran Point, Grand Canyon

If the result is a single answer and you've used voice search, your phone or tablet will speak the answer. If there are several results, you get a list to choose from.

Voice Actions

You can also use Google Voice commands to perform actions such as opening apps or setting an alarm. Some examples:

- Set alarm
- Send email
- Open Gmail
- Go to Google.com
- Weather.gov (opens weather.gov web page)
- Map of Yosemite National Park
- Directions to Sky Harbor Airport, Phoenix
- Navigate to nearest coffee bar

If Google Voice understands your request, it will perform the action. Otherwise, it will show a list of Google search results for you to choose from.

Google Now

Swipe up from the Home icon to display the Google Now screen. Google Now works with your Google account, non-Google apps that are connected to your Google account, and location information to provide you with timely information related to your current activities. This information is presented in the form of cards. For example, it learns your home and work locations as well as your commute route, and shows you traffic and commute times. It shows you the weather at home and your destination. You can customize Google Now by tapping the Menu icon at the left side of the Google Now search bar, but Google Now is really designed to customize itself based on your activities. The one setting you might want to change is Tablet Search. By default, Google Now searches Google apps such as email and calendar, but you might add apps such as Wikipedia, EverNote, and other reference and search apps (see Your Pocket Office.)

If any Google Now card is irrelevant, or you don't want to see it, just swipe it sideways to remove it.

A Google Search bar at the top of the Google Now screen lets you search Google by typing in a phrase or by speaking, just like the Google Search widget on the Home screens.

Google Now On Tap

New in Android Marshmallow, Google Now On Tap lets you bring up Google Now cards from nearly any screen. Just long-press the Home icon, and Google Now draws a box around the current screen. It then analyses the content of the screen and shows you relevant cards. For example, if you were texting a friend about a movie, Google Now will show you showtimes and theaters for that movie.

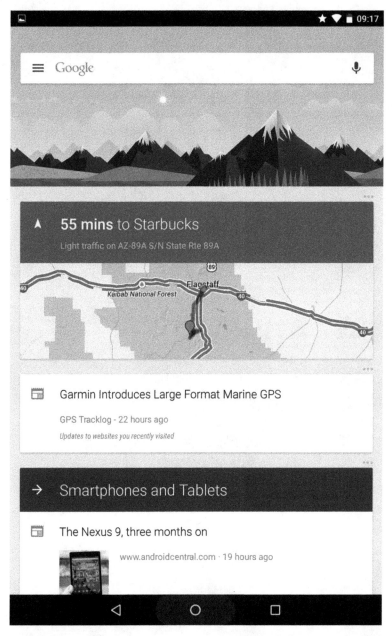

Google Now cards. Swipe up to see more cards, and swipe a card sideways to remove it

Browsing the Web

Google Chrome web browser comes pre-installed on Android devices. Since both Android and Chrome are Google products, it will come as no surprise that they work well together. However, many other browsers are available and have their own advantages, as you'll see below.

Google Chrome

If you have a Google account, all of your Chrome settings and bookmarks will be automatically synchronized with Chrome on your Android device. If you use other Google applications, such as Gmail and Calendar, their data is automatically synchronized as well. Because of the painless integration, Chrome will almost certainly be your primary browser on all Android devices you own, such as phones.

Chrome is available for all major platforms, including Android, iOS, Windows, Mac, and Linux. Changes you make to any Google app, such as Calendar, Contacts, etc., are automatically synchronized on your Android devices and desktop computers running Chrome. You can turn this integration off but it is one of the strong points of Android.

Tabbed Browsing

Chrome is a tabbed browser, which means you can have several web pages open at the same time in different tabs. The browser goes full screen and the tabs disappear while you're actively browsing a web page. To see the open tabs, swipe downward on the screen. To close a tab, tap the X at the right side of the tab. To open a new tab, tap the "New Tab" icon just to the right of the right-most tab.

New tabs open to your bookmarks. You can either select a bookmark, or type a web address directly into the address bar just below the tabs.

Bookmarks

The fact that Google Chrome synchronizes your bookmarks across all your devices that run Chrome means you can bookmark a web page anywhere and it will show up on your phone or tablet. Bookmarks are broken down into three folders:

Desktop Bookmarks

This shows the bookmarks on the Bookmark bar of Chrome if it's installed on a computer.

Other Bookmarks

This folder contains the rest of your desktop bookmarks, which are located in "Other Bookmarks" on your computer.

Mobile Bookmarks

This shows the bookmarks you've created on your Android phone and tablets.

Menu

The Chrome menu lets you open a new tab, open an incognito tab, go directly to your bookmarks, view your other Android devices, view and clear your browsing history, search within the current web page, and access Chrome settings and help.

Incognito Browsing

A feature of Chrome on the desktop as well as the Android app, Incognito Browsing lets you browse web pages without a record appearing in your browser or search history. Cookies downloaded to your device will be deleted when you exit all incognito tabs. Downloaded files and bookmarks you create while browsing incognito will be saved.

Incognito browsing is useful when shopping on line for presents, for example, and you don't want other family members to discover where you've been shopping.

As Chrome warns you, incognito browsing doesn't affect the behavior of other servers, computers, or people. In particular, if you log into your Google account while browsing incognito, your web history will be saved by Google in order to sync it to your other computers and Chrome devices.

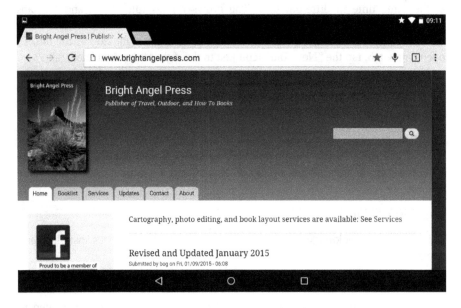

Google Chrome with two tabs. Web pages are usually designed for landscape orientation

Other Browsers

Some elements on web pages, such as drop down lists, may not work correctly on the Android Chrome app. In this case, you can try another browser. A search on Google Play reveals dozens of browsers, but here I'll focus on the most popular- Dolphin, Firefox, Maxthon, Puffin, and Opera (http://bit.ly/1dIFZVB).

Dolphin

One of Dolphin's strongest points is that it supports Flash video, as described below. It is also a tabbed browser that supports synchronized bookmarks and speed dial. Like Chrome, you can search from the address bar. Dolphin then lets you search YouTube, WikiPedia, Amazon, Ebay, or Twitter by tapping an icon at the bottom of the screen. There are many add-ons that extend Dolphin's capabilities.

Watching some videos embedded in web pages requires Adobe Flash Player, which is no longer supported by Adobe. You can still watch Flash video by manually installing an archived copy of the Flash Player app from the Adobe website, and then installing a browser that supports it, such as Dolphin. See Flash Video.

Dolphin Sonar is a voice search feature similar to Google Voice. It also lets you share on social networks, bookmark pages, and get directions.

Firefox

Firefox is a very popular browser that is available for all platforms. It also lets you sync across multiple devices.

One of Firefox's major strong points is the many add-ons available for both the Android and desktop versions. Adblock is one of the most popular add-ons, because as the name implies it lets you block and control those annoying pop-up ads. You can set Adblock's filtering level to let unobtrusive ads through, if you wish.

Firefox supports many different video formats but not Flash. It does support HTML5. A reader mode, accessed from an icon at the right end of the search bar, declutters web pages so you can concentrate on the text. Firefox is also one of the fastest browsers out there.

Maxthon

Like Chrome and Firefox, Maxthon lets you synchronize bookmarks and settings across multiple devices, including iOS devices such as the iPad and iPhone, Android, Mac, and PC. Unfortunately, Maxthon is not available for Linux computers.

Maxthon is a tabbed browser with the same appearance on all devices. It also features speed dial, which lets you put your favorite sites on the Quick Access Page. This page opens when you open a new tab, and contains small icons that you can tap to go directly to a favorite page. Maxthon is free and does not contain ads.

Puffin

Puffin supports Flash video right out of the box without having to install the Adobe Flash app. The browser also features a unique virtual trackpad which simulates a mouse and makes it easier to use some websites, especially those with drop down lists. The virtual trackpad is easily switched on and off by tapping a mouse icon at the lower left corner of the screen. There is also a virtual gamepad, accessed with an icon at the lower right corner of the screen.

Another great feature of Puffin is direct access to the Android virtual keyboard by tapping an icon at the lower right. Normally, the virtual keyboard appears automatically when you tap a text box or other screen element where text input is needed or allowed. But there are some websites where the keyboard doesn't appear when it should.

You can toggle the virtual keyboard between full screen, split, and compact formats. All three keyboards have forward and back icons, and copy, cut, and paste icons. To copy text from a web page or use it in a search, tap and hold anywhere on the screen to create a highlighted block. Drag the markers at the beginning and end of the selected text to select the desired text, then tap Copy or Search. The free version of Puffin limits you to 14 days of Flash viewing.

Opera

Probably best known for having pioneered the Speed Dial feature which lets you save your favorite websites as quick links, Opera also has a Discover page which lets you browse stories from major newspapers and magazines.

To access Speed Dial, open a new tab by tapping the tabs icon at the right side of the search bar, then tap New Tab. Swipe right to view your browsing history and left to see the Discover page. While on the History and Discover pages, swipe vertically to view more.

To add a website to the Speed Dial page, tap the + icon at the left side of the address bar. To open a new tab, tap the Tabs icon right of the search bar, and then tap the New Tab icon to open a regular tab, or the menu icon in the upper right corner of the screen to open a private tab. You can also close all tabs from the menu.

Off-Road mode lets you save data charges if you're on a mobile network (LTE devices only) instead of Wi-Fi by compressing data and saving on downloads. Turn on Off Road mode by tapping the "O" menu icon in the upper right corner and tapping Off Road. Tap the "O" at any time to see your data savings, displayed below the Off Road toggle.

Keeping Up with Social Networks

There are Android apps for all the major social networks, so you can use your device to keep up with your friends and social feeds just as you can on your phone or computer.

Flipboard

Flipboard (http://bit.ly/14VarZ7) is an alternative to using separate apps for each social network. You can link Flipboard to Google Plus, Facebook, Twitter, LinkedIn, and many others, which lets you browse postings from many different social networks in one elegant app.

Flipboard can also be linked to news sources and RSS news feeds. For more information, see Keeping up with the News.

Google Plus

It probably won't be a surprise that the Google Plus app comes pre-installed on your Android device. As soon as you log in to your Google account from any Google app, Google Plus is synchronized with the other devices where you also use Google Plus.

Flipboard home screen showing several feeds. Tap the Refresh icon at the lower left to refresh all feeds, and tap a feed to open it

The Google Plus app for Android offers a clean and fast user interface, which makes it easy to keep up with your friends and also join Google Plus Communities. The ease with which you can join groups of people with shared interests is one of the things that sets Google Plus apart from Facebook.

One of the unique features of Google Plus is Hangouts, which makes it very easy to start a group discussion via the keyboard or video call. Another nice feature of Google Plus is that you can view photos that you've uploaded to Google's Picasa photo sharing service.

Facebook

Facebook has a free Android app which you can install, but there are plenty of alternatives (http://bit.ly/1biMQmP), including Flipboard which I mentioned at the start of this chapter.

HootSuite (http://bit.ly/1dQytHl) lets you manage Facebook, Twitter, LinkedIn, and many other social networks from the same app. HootSuite also has a flexible web page that you can use to manage multiple social networks. The free app is limited to five social networks, which includes Facebook groups and pages- to manage more than that, you'll have to upgrade to HootSuite Pro for $9.95 per month. If you have more than five social networks to manage, I suggest using Flipboard instead, which is free and has no limits.

Twitter

The Twitter app for Android (http://bit.ly/1aCrXSC) is free and works very well. For an alternative, see Flipboard.

Instagram

Instagram (http://bit.ly/199plye) also has a free app, and you can use Flipboard as an alternative. There are many add-on apps for Instagram (http://bit.ly/17fojyc) that let you merge Instagram photos, add frames, format text, manage followers, and more.

Pinterest

As with the other popular social networks, Pinterest offers a free app (http://bit.ly/18h1eup). As of yet, Flipboard doesn't allow you to connect your Pinterest account, but there are many other apps that enhance Pinterest.

An example is PinHog (http://bit.ly/19hd1pP). This app extends Pinterest by letting you browse all categories with a channel scroll bar, make pins available offline, set pin images as wallpaper, pin web pages from web browsers, share pin link to Twitter, Facebook, and email, and search for downloaded pins with keywords.

Vine

Vine is the hottest new social networking app (http://bit.ly/1fU4dNB). The video equivalent of Instagram, Vine lets you create and share short looping videos on Vine as well as on Facebook and Twitter.

LinkedIn

LinkedIn, the social network for professionals, has a free app, http://bit.ly/19Nm5VE, and you can also connect to your LinkedIn account with Flipboard.

Make Video Calls with Skype

Skype (http://bit.ly/18Krm1Q) lets you do free instant messaging, voice, and video calls with any other Skype user as long as you're both on the Internet. You can also call cell phones and landlines and message cell phones at low cost. And you can do group conference calls- free between Skype devices, and for a per-minute charge if landlines or cell phone are in the conference. All you need is a free Skype account, and you can sign up using the Skype app on your device or at skype.com on your computer.

For a monthly fee, you can upgrade to a premium account, which adds group video calls, group screen sharing, removes ads, lets you access live chat customer support, and make unlimited calls to landlines or cell phones in a country or region of your choice.

As Skype warns you when you set up your account, you should not use Skype to make emergency calls. Skype is not set up to make calls to emergency numbers such as 911. Always use a cell phone or landline for emergency calls.

Setting up a Skype Account

When you start Skype for the first time, you'll see the sign in screen. To set up a new account, tap "Create a Skype account". After agreeing to the Terms of Use, you're taken to the "Create account" screen. Enter your name, Skype name, password, email, and phone number with country code.

If your chosen Skype name is available, your account will be created. Otherwise, you'll be shown a list of alternatives and asked to either cancel or choose one of the alternatives.

Calling a Skype User

Skype opens to your contact list. To call another Skype user, you'll need their Skype user name. If you don't know the user name, you can tap the Search icon to look it up in your contacts or in the Skype Directory.

Once you've found the user name in the Skype Directory, you have to add it to your Skype contacts before you can make a Skype call.

Small icons show the status of each contact:

- Online (green): Contact is online and available to talk

- Away (yellow): Contact is online but away from Skype so they might not pick up

- Busy (red): The contact is online but busy

- Offline (white): The contact is offline

- Landline (phone icon): The contact is on a cell phone or landline. You'll need pay as you go credit or a subscription to call them

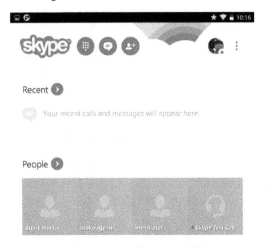

Contacts list in Skype

- Pending (? icon): Contact hasn't yet accepted your contact request
- Blocked (circle/slash icon): A contact you have blocked

To call a Skype contact, tap the contact to open it, then tap the Phone icon at the upper right corner of the screen. If the contact accepts the call, you'll be connected and can start talking.

During a call, you can mute your microphone at any time by tapping the microphone icon. When you're finished with your call, tap the red phone icon to hang up.

Answering a Call

Skype starts the Skype call screen and notifies you with tones when a call comes in. If your device is off, it will turn on and show the Skype call screen. To answer the call, tap the green phone icon. To decline the call, tap the red phone icon.

If you have a lock screen enabled and your device is locked when a call is received, you can still answer the call. To see your call history and access other Skype features, tap the Unlock button on the left. This takes you to your unlock screen. After you unlock the screen, it returns to the Skype call screen. When you're finished with your call, tap the red phone icon to hang up.

Group Calls

You can't start a group call from your tablet or phone, but you can join a group call. Skype will notify you that you have an incoming call. Just answer it in the usual way, by tapping the green phone icon.

Making a Video Call

Select the contact to call, then tap the video icon at the upper left. Note that the person you're calling may have video turned off by default. In this case you won't see the other person.

During your video call, you can turn off video or switch to the rear camera by tapping the video icon at the bottom of the screen. As with a voice call, you can mute the microphone by tapping the microphone icon.

Sending Instant Messages

To send an instant message (IM), select the desired contact. Then type your message in the instant message box at the bottom of the screen. You'll see the response, if any, below your message.

Sharing Files

You can also send files of any type and size to your Skype contacts. To send a file, select a contact, then tap the + icon at the upper right corner of the screen. Select Send File. You'll then be asked to choose the app that you want to use to select the file.

Note that the recipient will be asked whether to accept the file. If they decline, then sending will fail. You can also send files during voice and video calls by tapping the + icon at the bottom of the screen.

Calling Cell Phones and Landlines

You can use Skype to call cellular phones and landline numbers by either paying as you go, by the minute, or by paying for a subscription. If you only call occasionally, paying by the minute is the better option. However, if you make a lot of calls to the same countries or regions, a monthly subscription is cheaper.

For example, as of this writing, calls to the 48 United States phones cost 2.3 cents per minute plus a 4.9 cent connection charge per call. In contrast, by subscribing, you can make unlimited calls to the 48 United States and Canada for $2.99 per month. So if you call for more than 130 minutes per month, the subscription is cheaper.

Pay as you go rates vary greatly by country and even within a country, depending on the phone service that owns the number you're calling. If you do a lot of worldwide calling to cell phones and landlines, you may want to consider an unlimited world subscription, which is currently $13.99 per month.

Note that you can call landline phones in over 60 countries at present. The list of countries where you can call cell phones is much smaller- currently the United States, Canada, United Kingdom, China, Hong Kong, Guam, Puerto Rico, Singapore, and Thailand.

Setting Your Status

Skype automatically changes your status to Online when you start Skype and to Offline when you exit Skype or turn the device off. If you haven't been active on your tablet for 5 minutes, Skype will set your status to Away.

You can manually set your status to Online, Offline, and Away. In addition, you can set your status to Invisible and Do Not Disturb. To set your status, tap Profile on the left side of the screen. Then tap the desired status to set it. Skype will notify you of an incoming call if your status is set to anything except Offline. Invisible status makes you appear offline to all of your contacts, but you can still use Skype normally. Offline status disables Skype- you can't make or receive calls, or send or receive instant messages or files.

Checking Email

Android devices usually come with two email apps, a general email app and one specifically for Google's Gmail. However, there are plenty of other email apps you can install.

Gmail

If you already have a Google account, you will have signed into it when you first set up the new device. To use Gmail simply open the Gmail app. The app gives you most of the power of desktop web-based Gmail, with the notable exception being that you can't send email blasts to groups that you've created in Gmail Contacts.

However, there are plenty of apps that let you do group emails. One such app is DW Contacts (http://bit.ly/193RkM6). Although designed for Android phones, it works well on tablets as well. To send an email blast (assuming you have one or more Gmail groups already set up), open DW Contacts, then tap Groups at the bottom of the screen. Tap the group you wish to email, then tap the menu icon at the upper right corner of the screen. Select "Send email to shown contacts". You can tap to uncheck anyone you don't wish to include in this email. At the top of the screen, you can select whether the address will be in the To, CC (copy to), or BCC (blind copy to) address fields. When finished, tap OK, then select the Gmail app. Compose your email, then tap the Send icon at the top right of the screen.

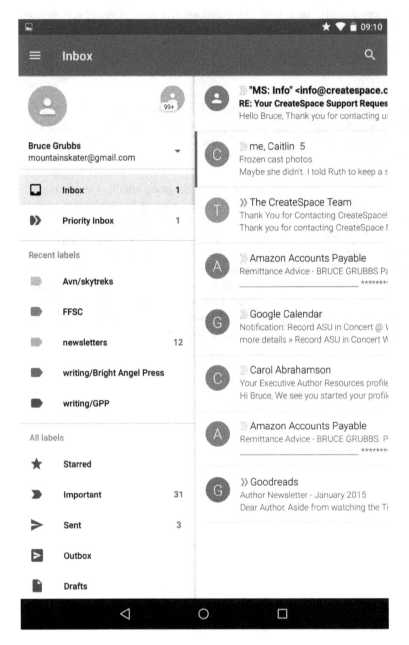

Gmail showing labels on the left and email headers on the right. Tap any email to open it

Other Email Accounts

The generic Email app can be used with email from most Internet service providers as well as web mail providers such as Gmail, Hotmail, and Yahoo Mail.

When you first open the Email app, it walks you through setting up an email account. At a minimum, you'll need your email address and password. For many email providers, that's all the information you'll need. For others, you may need the POP3 or IMAP server address, incoming port number, SMTP server, security settings, and outgoing port number. You'll need to obtain this information from your email service provider. You can set up multiple email accounts with the Email app.

Managing Contacts

If you use Google's Gmail to manage your contacts, all of your Gmail contacts will be available when you sign into your Google account during device setup. And as with all other Google applications, your contacts will be synchronized between all your Android devices (such as a phone) and Google applications on the desktop web versions. This means you can add or update a contact one Android device and the changes will show up on your desktop computer and on your Android phone.

Contact Apps

The built-in contact app, Contacts, is good enough for basic contact management, but you can add a lot more power with apps from the Google Play Store.

With the Contacts app, you can add, edit, and delete contacts and groups. You can also create favorite contacts, and look at your most frequently contacted contacts.

DW Contacts

As mentioned above, other apps let you do a lot more with your contacts. DW Contacts (http://bit.ly/193RkM6) is a powerful contact manager with many options. You can create, copy, edit, change, and delete contacts either singly or using multi-select. Most operations can be selected from the menu icon or by long-pressing a contact, group name, or item. You can create appointments in your Google calendar directly from DW Contacts by long-pressing a contact.

Contacts +

If you want a simple, clean interface, Contacts + (http://bit.ly/1ggAcYF), offers many ways to sort and view your contacts. It can synchronize with Google+ and Facebook to display your contacts images in three different sizes in a grid view. If most of your contacts don't have photos, you can switch to list view. You can sort alphabetically by first or last name, or by frequency or last used.

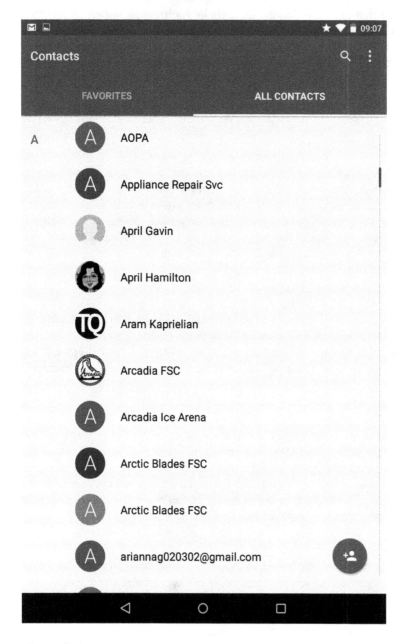

The stock Contacts app

Listening to Music and Radio

There are many ways to listen to music and radio on Android phones and tablets. Of course, Android integrates tightly with the music store on Google Play, but you can install apps that let you listen to music that you've bought from Amazon and music that you store on the device and listen to offline.

Music can be stored on your device so you can listen to it when you're not connected to the Internet. If you have a large music collection and often listen in areas without Wi-Fi, I suggest you buy a device with as much internal memory as you can, or that has an SD or micro-SD slot for adding memory. Of course, if you have an LTE tablet or a phone, you can listen online anywhere the device can connect to the cellular data network- but you will pay for data used on your plan.

All of the music apps that I describe below let you sort through your music by album, artist, and song, as well as create custom playlists. They also have an equalizer mode so you can fine-tune music playback for types of music, listening devices such as earphones and speakers, and your particular listening environment.

Controlling the Volume

Most Android devices have physical volume up and down buttons. These buttons work even when the screen is locked or asleep. If the screen is awake, a volume slider appears and shows the volume level. You can also tap or slide the slider to change the volume, or tap the drop-down at the right to select other volume options.

Google Play Music

Google Play Music app comes installed on Android and lets you play music that is stored on the device as well as music that you've bought from Google Play and is stored on Google Drive.

If you have a music collection stored on a computer, you can easily add it to Google Drive so that it's accessible from Play Music on all your devices, including Android phones and on computers. To do this go to Google Play Music on your computer's web browser, then My Library. Download the Google Play Music Manager (http://bit.ly/19tBwAh) for your platform- it's available for Windows, Mac, and Linux- and run it. After you specify the location of your music on your computer, Music Manager will upload your music to Google Drive in the background. You can also specify that any new music added to your computer's music folder(s) will be automatically uploaded in the future.

Google provides space for up to 20,000 songs. It does a smart upload- if a song is already in the Music Play Store, it links to that copy rather than uploading from your computer. If you have a large collection, that can save a lot of time.

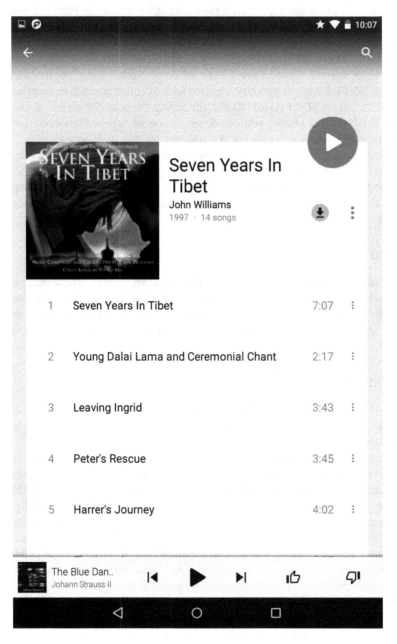

Google Play Music

If you turn your computer off before the upload is complete, it will resume the next time you turn your computer on.

Of course, music in the Google Cloud can only be played when your Nexus is connected to the Internet via Wi-Fi or LTE. To download music to the Nexus for play offline, tap the Keep icon (a pushpin) on the album or song's detail page. See the status of music downloads, tap the Menu icon in the upper right corner, then Settings, and then View download queue.

Google Play All Access (http://bit.ly/GFTs4n) is a subscription service, currently $9.95 per month, that gives you unlimited access to millions of songs and allows you to create personalized radio from any song or artist.

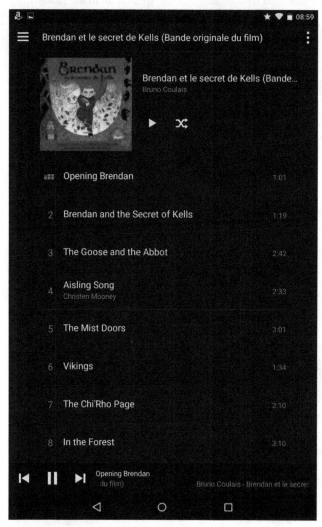

Amazon MP3 player

If enabled, the equalizer mode lets you control the frequency bands of its built-in presets as well as create your own user settings.

Amazon MP3

Similar to Google Music Play, Amazon's MP3 app (http://bit.ly/1ycBdgb) gives you access to music you've bought from Amazon's MP3 Music Store. You can play music stored on the device as well as in the Amazon Cloud, and you can play it from any Android device, Kindle Fire tablets, and computers.

To download an album or song, tap the Download icon (a down arrow) on the detail page. To check the status of downloads, tap the Menu icon in the lower right

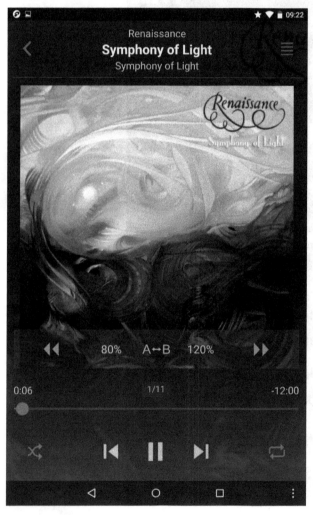

Jet Audio playing

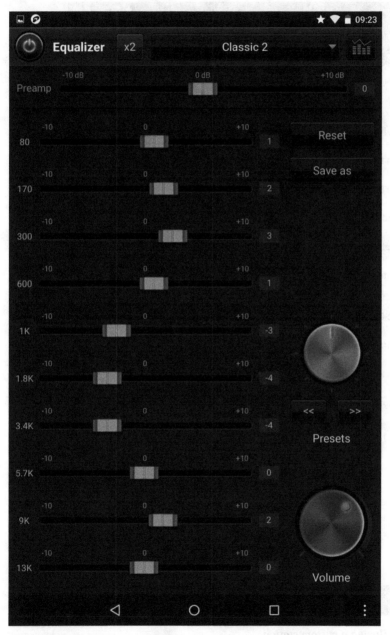

The graphic equalizer in Jet Audio

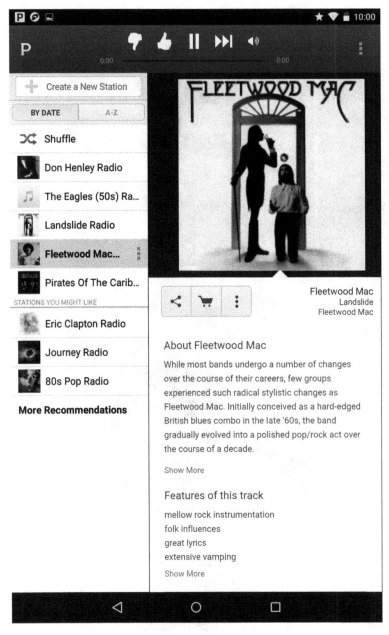

Pandora playing a custom station

corner of the screen and then tap Downloads. Amazon MP3 equalizer mode is limited to built-in presets.

JetAudio

A paid app, JetAudio Music Player Plus (http://bit.ly/15Ukf2P) only plays the music stored on your device, but it does it very well. JetAudio has an extremely capable and customizable graphic equalizer which can be set to control 10 or 20 frequency bands. It has a large number of presets for different music plus four user presets that you can customize. It also has four sound effect presets where you can control and save extreme bass, extreme wide, hall, AGC, and pitch levels.

The JetAudio interface is also customizable with skins and many other features. You can sort music by the usual artists, albums, and songs, create custom playlists, and sort music by folder and genre. You can try JetAudio Basic (http://bit.ly/GGqlxa) for free. This version has ads and a few features are disabled.

Pandora

The free Pandora app (http://bit.ly/1faJoia) connects you to Pandora's well-known Internet "radio" service. Pandora lets you create a custom "radio station" by searching for a favorite song, artist, or genre. It then creates a custom playlist. As Pandora plays additional music, you can give each song a thumbs up or thumbs down. Pandora uses this information to customize your radio station.

Pandora requires that you create a free account so it can track your preferences. With a free account, you'll see and hear ads. Serious Pandora users will probably want to subscribe to Pandora One, currently $3.99 per month. Pandora One removes the ads and provides a higher quality audio stream.

TuneIn Radio

TuneIn Radio (http://bit.ly/195KZl9) is a free app that lets you stream more than 70,000 radio stations from around the world, plus millions of podcasts, shows, and concerts. The paid Pro version (http://bit.ly/19sgBn4) adds the ability to record everything you hear on TuneIn. You can also set up timed recordings.

You can sign in with your Google account or create a TuneIn account. This lets you listen to your TuneIn favorites from any device or computer. On a computer, go to http://tunein.com.

Streaming on the Go

If you have an LTE device, you can stream music, Internet radio, and radio stations from anywhere that you have a cell phone data connection. Although you will use data from your data plan's allowance, audio files are much smaller than video files so you can listen to a lot of music without paying excess data charges.

Most cell phone companies will allow you to set up alerts so you'll get an email or text when you approach your data limits.

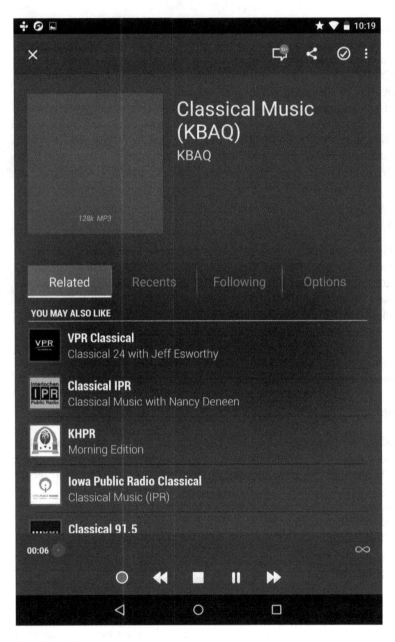

TuneIn Radio

Playing Games

An entire book could be written about this subject, so I'm just going to touch on a few games here that show off the graphics and speed of Android.

One caution- most free games are ad-supported, and many only let you play for a limited time or to a limited level before requiring you to buy the paid version.

Here are a few classic games to get you started.

Angry Birds

The hit from Roxio now comes in many versions, both paid and free, but the original (http://bit.ly/GFYe1J) is always a good place to start. The birds are steamed because the bad pigs have stolen their eggs, so your job is to fire different types of angry birds at the structures protecting the pigs and destroy them.

Random Mahjong

This free app (http://bit.ly/17akHdN) is a beautifully rendered version of the classic tile-matching game. You can play at levels from 32 to 152 tiles as well as customized levels. Games are saved if you quit before finishing, and you have undo, rewind, and shuffle to make games even more fun. The free version is ad-supported.

Chess Free

There are many free Android chess apps but this one stands out (http://bit.ly/196hnlI). There are ten play levels from novice to expert. You can watch the computer's thinking as it plays to help you learn the game. There are four different boards and four different chess piece sets. This version is ad-supported. The paid version (http://bit.ly/16MVID5) eliminates the ads.

The developer, AI Factory Unlimited (http://bit.ly/1adivUJ), has many of the classic board and card games, including Sudoku, Backgammon, Reversi, Checkers, Spades, Gin Rummy, Hearts, Solitaire, and Go.

Simple Rockets

This physics-based game (http://bit.ly/18PWQji) accurately models orbital motion using the equations developed by Kepler 400 years ago. You build multi-stage rockets from a list of parts, then launch them from any planet in the solar system. You can go into orbit and attempt to land by parachute or rocket on planets with and without an atmosphere.

Running Android

Random Mahjong

60

Flying a multistage rocket into orbit in SimpleRockets

More Games

Remember that it's easy to try the free versions of games. So just browse the games category on Google Play (http://bit.ly/15Q79c7) and have at it!

Just don't forget to uninstall any games you don't like to avoid using up the space on your device. To uninstall any game or app, find it in the Apps Drawer, long-press its icon, then drag it to the Uninstall area at the top of the screen. When the icon turns red, release to uninstall it.

Taking Pictures

Most Android devices have both front and rear-facing cameras. The front camera is usually lower-resolution and fixed focus, mainly suited for Skype and other video apps, as well as taking selfies for social media sites such as Facebook and Instagram. The rear camera is generally higher resolution with auto focus. There are photographer's apps that turn your device into a valuable tool for serious amateur as well as professional photographers. The main limitations of Android phone and tablet cameras are that the sensors are very small and don't perform well in low light, and the lenses are wide-angle, which creates unflattering photos of faces (the big nose syndrome) and creates distortion in scenic shots. Phone cameras have one huge advantage, though, even for you serious amateur and professional photographers- it's always with you.

Google Plus Photo Backups

By default, Google Plus automatically backs up your photos to Google Drive. Though these photos are private unless you enable sharing, you may wish to disable this setting. To do this, open the built-in Google Plus app, tap the Menu icon, and tap Auto Backup. Tap the On/Off icon at the upper right corner to turn auto backup off.

Camera Apps

Camera

The built-in Camera app does the basics and may be all you need. It has panorama, sphere, video, and still modes. You can select between the front and rear cameras and set exposure compensation. A "More Options" menu lets you turn on location recording, use a self-timer to take delayed pictures, and set the image size. White balance can be set to auto, daylight, fluorescent, tungsten, or cloudy. You can also choose from four Scene Modes- action, night, sunset, and party. Also, you can activate the Camera app by rocking the phone, even when it is locked.

ProCapture

If you want more capability in your camera app, have a look at ProCapture (http://bit.ly/16L3xH0). The free app offers everything the paid app has except higher resolutions and tap-to-focus.

While some of the features may seem like overkill for a 5mp camera, don't forget that you can experiment and take test shots, even if you do most of your shooting with a higher resolution point-and-shoot or single lens reflex (SLR) camera.

Running Android

Icons along the side of the screen give you quick access to the most commonly used functions. Normal, timer, burst, reduced noise, wide shot, and panorama shooting modes are supported. Burst allows you to take several shots in quick succession- useful for action. Reduced noise reduces the digital noise in your shot, which tends to be worst in low-light situations. WideShot links together three images to create a wide angle photo. Panorama stitches together up to 12 photos to create a panoramic shot.

Focus modes include auto, continuous, infinity, and macro. There are 14 scene modes which preset the camera for a wide variety of shooting conditions. A large number of color effects let you take images in black and white, sepia, and many other modes.

You can turn location storing on and off, and set white balance to auto, incandescent, daylight, fluorescent, cloudy, shade, twilight, and warm fluorescent. Unfortunately there's no custom white balance, which is sometimes necessary to deal with fluorescent lighting. Exposure compensation of up to plus or minus 2 stops is settable in tiny increments.

You can turn on an exposure histogram, as well as two composition aids- a Golden Mean grid or a Fibonacci Spiral. You can also choose from several aspect ratios and set the image size and image quality.

From the menu icon, you can open photos for viewing using the built-in Photos app or use any other viewing app you've installed, such as QuickPic. You can also switch cameras. Contrary to the description in the Google Play Store, ProCapture can take videos- it just switches to the built-in Camera app to do so. A Settings menu gives access to burst and timer modes as well as general settings.

Pro HDR Camera

This paid app (http://bit.ly/1y2As7k) creates a High Dynamic Range (HDR) image by shooting an exposure for the highlights and another for the shadows, then blending the two to create an image with a greater range of tonal values than the camera can capture.

As experienced photographers know, no camera sensor has a wide enough dynamic range to capture what the human eye sees in high contrast lighting such as a sunny day with deep shadows. HDR effectively expands the dynamic range of the camera so that you can capture details in the shadows without blowing out the highlights.

Photo Viewing

Photos App

The Photos app is pre-installed on Android. Like the Camera app, it gives you the basics. You can view images in various locations as thumbnails and full screen and run a slide show. The Photos app goes a bit further- it connects to any photos you've stored on Google Picaso.

QuickPic

QuickPic (http://bit.ly/1cqr5Sw) adds a lot more photo viewing capability, and it's free and has no ads. You can specify the folders to be scanned for images and exclude folders to speed up the scan. You can hide images and password protect them. Hiding is especially useful if you have a lot of music stored on your Nexus. You can use the hide feature to hide album cover art so that it doesn't clutter up your photo collections.

QuickPic can also play animated GIF's and videos, although the movie playback is basic. You can rotate and resize images as well as set them as wallpaper. File management features allow you to rename, sort, move, and copy images, and you can create new folders to help you organize your photos from within the app.

Photo Editing

Most Android owners will be interested in photo editing for fun, such as adding special effects, rather than pro-level editing, because of the limitations of the cameras. But there are editing apps available to suit almost any level of photo editing.

Snapseed

A free app, Snapseed (http://bit.ly/173E8La) is an easy to use yet powerful photo editor with a beautiful interface. After you take a new photo or open an existing one, a scrolling series of large tool icons appear along the bottom of the screen. The first time you select a tool, a help overlay shows you how to use the tool. Later, if you forget how to use a particular tool, you can bring up the help overlay by tapping the question mark icon at the upper right corner.

Despite the simple interface, Snapseed offers a wide range of tools, including auto correct, selective adjust, tune image for brightness, ambiance, contrast, saturation, and white balance, straighten and rotate, crop, details (sharpening), black and white conversion, vintage film effects, center focus, tilt-shift, retro effects, and frames.

Adobe Photoshop Touch

For professionals or advanced amateurs used to the full control provided by powerful desktop photo editors such as Adobe Photoshop and Gimp, this paid app (http://bit.ly/17uEa98) has many of the controls of desktop editors, such as layers, filters, selection tools, and adjustments. It adds features unique to the tablet environment, such as camera fill. This lets you fill an area in a layer with an image captured by the camera.

Google Image Search is integrated, which allows you to easily search for images on the Internet and edit them. Projects can be synchronized with the Photoshop desktop versions CS 5.1 or CS 6.0.

Photos app

Editing a photo in Snapseed

Photo Sharing

See the chapter *Keeping up with Social Networks.*

Photographer's Aids

There are many photographer aids on Google Play, such as light meters and time lapse calculators, but one really stands out:

The Photographer's Ephemeris

The Photographer's Ephemeris (http://bit.ly/19N4vTE) is a powerful assistant for the outdoor and landscape shooter. Essentially, it is a map-based sun and moon calculator which lets you see how the light lays on the land nearly anywhere in the world, now or in the future. Sun and moon calculations do not require an Internet connection.

TPE automatically detects time zone and elevation and corrects for atmospheric refraction and height above horizon. You can also determine when the sun or moon will rise or set behind hills or mountains.

Among the many features that are displayed for the current location are time and direction of sun and moon rise and set, phase of the moon and percent illumination, times of civil, astronomical, and nautical twilight, current azimuth and elevation of the sun and moon, distance, azimuth, elevation angle of a selected point on the map, and the elevation and distance of the horizon along a selected azimuth.

TPE opens to a map view, centered on your last location. To center on your current location, tap the Center icon in the upper right corner. As with other location services, TPE location uses Wi-Fi, the cellular network (on LTE models), and GPS to find your location. GPS is by far the most accurate but since it depends on orbiting satellites it may not work inside buildings or in narrow canyons where the view of the sky is limited or blocked. To select another location, you can either search for it by name or drag the map pin.

You can select from standard (street view), satellite, hybrid, and terrain maps. Note that in order to display maps, the phone or tablet has to have an Internet connection, either via Wi-Fi or LTE. If you have a Wi-Fi-only tablet, plan on checking the map from an indoor location before you head to the field. Since I also have an Android phone, I use my Wi-Fi Nexus for planning outdoor photo shoots in advance from home or a place with Wi-Fi to take advantage of the larger screen on the tablet, then use TPE on my phone in the field.

The usual Android gestures work on the TPE maps to pinch rotate, drag pan, and pinch zoom in and out. You can also enable auto-compass mode so that the map will automatically orient itself to the direction you're facing.

If the time is displayed incorrectly or you get an error when searching for a location, do a soft reset of the phone or tablet.

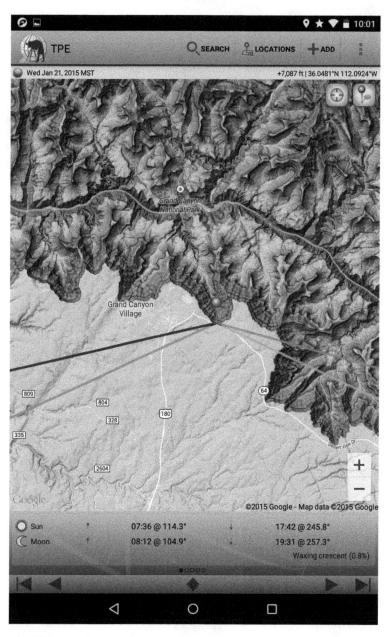

The Photographer's Ephemeris

Watching Movies and TV

On Google Play, you can choose from thousands of movies and TV shows to watch. Movies and shows are available to buy or rent, depending on the specific video. Videos that you buy are saved to your Google Play library and are available for streaming to your phone, tablet or a computer for viewing any time. Videos that you rent are available for varying amounts of time from the time of purchase and after you start watching the video. You can review the rental period during the payment process.

Videos can also be downloaded to a phone or tablet (but not to a computer) for offline viewing when you don't have an Internet connection, such as on an airplane. Keep in mind that videos use very large amounts of memory so you won't be able to store very many on your device unless it has an SD or micro-SD memory card slot. For example, a 90-minute feature film in HD uses about 1.4GB of storage.

Shopping for Movies and TV Shows

You can shop from the device or on a computer. On the computer, go to play.google.com and then click on Movies and TV. On your phone or tablet, open the Play Store app or the Play Movies & TV app. The shopping experience is very similar on Android and on a computer.

You can search for a movie, browse new movie and TV releases, top movies and TV, and browse by genre.

Once you've found an interesting video, tap it to see the detail page. On the detail page, you can save the video to your wishlist, share it with others, see the cost to buy or rent, view trailers, read the synopsis, and read reviews. There is also information on the cast and credits and the run time, as well as the rental period.

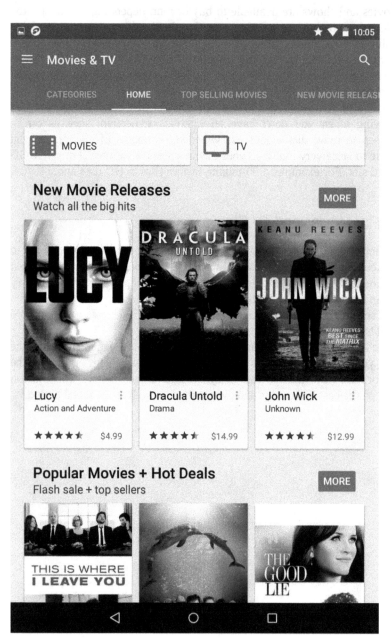

Google Play Movies store

Watching a Movie or TV Show

Once you've bought or rented a movie or TV show, you can watch it on your tablet, phone, or computer.

Watching Online

Since streaming video- especially HD video- uses a lot of data, it's best to watch when you have a strong Wi-Fi connection. You can watch streaming video over LTE but you'll use gigabytes (GB) of data.

You can quickly go through your monthly data allowance and then pay high rates for the excess. Most cell phone companies allow you to set up alerts so you'll get an email or text when you approach your data limits.

Watching on your Tablet or Phone

To watch a movie or TV show that you've already bought or rented, open the Google Play or Play Movies app and then tap on My Movies & TV. (If that choice is not visible, tap the Menu icon in the upper left corner of the screen to display the menu.) Browse or search for the video you'd like to watch, then tap to play it.

Video automatically fills the screen. Tap the screen to show the controls. To return to the previous screen, tap the Back icon at the upper left corner of the screen. A Play/Pause icon appears at the lower left corner, next to the progress bar. You can fast forward or jump to a different place by tapping the progress bar or dragging the slider. The run time and total run time are also shown. In the lower right corner, icons let you switch between SD (Standard Definition) and HD (High Definition) video. You can also turn captioning on and off. Tap the Menu icon in the upper right corner to access settings and help.

Watching on a Computer

To watch a movie or TV show you've previously bought or rented, open your web browser and go to play.google.com/movies. Or from Google.com, click on the Apps menu at the upper right corner of the Google window and select Google Play. Now select Movies & TV from the menu at the left side of the screen, then My Movies & TV. Browse or search for the video you'd like to watch, and click on it to open it.

Controls and a progress bar at the bottom of the screen let you play and pause, adjust the volume, fast-forward or jump to a different place, change video resolution, and switch to full screen. You can exit from full screen mode by pressing the ESC key at any time.

Downloading a Video to Watch Offline

As mentioned earlier, you can download movies and TV shows and that you've bought or rented to your phone or tablet and watch them when you don't have an Internet connection. To do this, find the video you wish to download in My Movies & TV on the Play Movies app, then tap the small gray Pin icon to start the

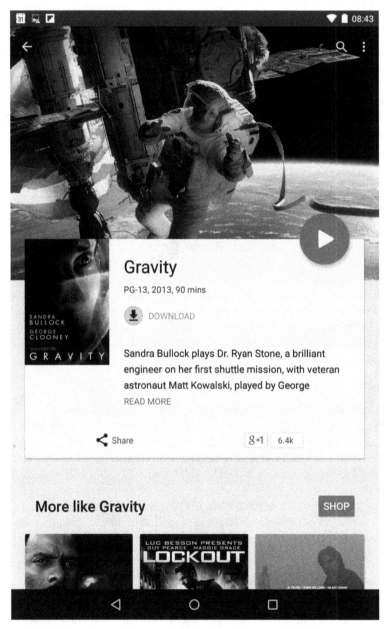

To stream this movie, tap the large play icon. To download it to watch offline, touch the download icon

download. During the download, the Pin icon starts to fill in with red. You can check the download progress by swiping down from the notification area at the upper left corner of the screen.

Once a video has been downloaded, the Pin icon turns red. To delete a video or stop the download, tap the Pin icon and confirm that you wish to stop or delete.

Netflix

If you have a Netflix account, you can watch streamed movies and TV shows from Netflix. All you need to do is install the free Netflix app (http://bit.ly/1aPNDy1).

Flash Video

Flash video is no longer supported by Adobe and the Chrome browser. The easiest way to watch flash video on Android is to install Photon Flash Player and Browser (http://bit.ly/1C5sPhL). Unfortunately the free version overlays part of the screen with ads, and constantly nags you to upgrade to the premium version for $9.95 per year.

Puffin (http://bit.ly/1cy3mTZ) is another alternative. The free version lets you watch Flash video for a 14 day trial period. To continue watching Flash after that you'll have to upgrade to the paid version (currently $2.99). Unlike Photon Flash Player, this is a one-time charge. I suggest you try Puffin to see if it meets your needs.

Yet another way to enable Flash is to install the Adobe Flash app and a supported browser. This is an unauthorized app, so to install it, open Settings, then scroll down and open Security, and turn on Unknown Sources.

On your device, open a browser and go to this link: http://adobe.ly/173UyBP. Scroll down to "Archived Flash Player versions for developers" and continue down to "Flash Player for Android 4.0 archive". Tap on the latest version at the top of the list to download it. Watch the download icon at the top left of your screen- when the animation stops, the file has been downloaded. Open the notifications window, then tap on the Adobe Flash Player line to start the install.

After the install is complete, go back to Settings, Security, and uncheck Unknown Sources.

Next, install the Maxthon web browser from Google Play: http://bit.ly/16IPU9F. This seems to be the best browser app for Flash on Android.

Now, open Maxthon and browse to the desired website. The video window displays the standard Flash video controls.

Reading Books and Magazines

With your Android tablet or phone, you have access to millions of books and thousands of magazines through Google Play, Amazon Kindle, Kobo, Project Gutenberg, libraries, and other sources. In addition, you can rent textbooks through Google Play.

Google Play Books

As with everything else in the Google Play store, you can shop on the computer at play.google.com/store/books or from the Play Store and Play Books apps. From the Books home page, you can search for books or browse by category or genre. Categories include new arrivals in fiction and nonfiction, top selling, and top free.

When you find a book that looks interesting, tap it to open the detail page. On the detail page, you can add the book to your wishlist, share with friends, try a free sample, and buy the book. You'll also see the publisher and publication date, the length of the book, the publisher's description, reader reviews, and about the author.

Take advantage of the free sample, especially with authors you don't know. There are no strings attached and reading the free sample gives you a chance to see if you like the book before spending any money. It's the electronic book equivalent of browsing in a physical library or bookstore.

To read a book that you've bought, go to My Library in the Play Store or Play Books app and tap on the book to open it. Books always open at the last place read.

Reading Books

By default, books display full screen- one page in portrait orientation and two pages in landscape view. Touch the middle of the screen to bring up the menu bar at the top of the screen and the progress bar at the bottom.

To go back to My Books, tap the back arrow and book title at the upper left corner of the screen, or the Back icon at the bottom of the screen.

Searching

You can search within the book by tapping the search bar at the top of the screen and typing in a search term. Tap the Search key on the virtual keyboard to start the search. Tap any of the search results to go to that page. You can continue to jump to result pages by tapping the forward and back arrows at the bottom of the screen. To return to the place you were reading, tap Back until you've returned to My Books, then tap the book to reopen it.

Google Play Books library

Reading a book in Play Books with the Menu and Progress bars shown

Changing Fonts and Size

Tap the Aa icon at the top of the screen to change text settings. You can change themes (backgrounds), the typeface, text alignment, screen brightness, font size, and line height (spacing).

Menu options

Tap the Menu icon at the upper right corner to see the menu. You can change to the original pages view, tap About this book to see the book description in the Play Store, share the book, add a book mark, turn on reading aloud, access settings, and get help.

Using the Table of Contents

Tap the center of the screen to bring up the progress bar and then tap the Contents icon at the lower left corner. Tap any entry to jump to that chapter in the book.

The Progress Bar

When you tap the center of the screen while reading, the progress bar shows your current place in the book. Tap anywhere on the progress bar or drag the slider, to jump elsewhere in the book. Your current page number and the total pages are shown in the lower right corner.

Bookmarks

You can set bookmarks from the menu, but the easiest way is to tap the upper right corner of the screen. A red bookmark appears.

To view your bookmarks, tap the Contents icon at the lower left corner of the screen (tap the center of the screen to bring up the progress bar if it's not visible.) Then tap the Bookmarks tab.

Notes and Highlighting

Long-press in the text to select it. Then drag the markers to select the desired text. Tap the highlight icon to highlight the text and choose from four highlight colors. To create a note, tap the Notes icon, then enter your text with the virtual keyboard. A small Note icon appears next to the highlighted text.

To view your notes and highlights, tap the center of the screen, then tap the Contents icon at the lower left. Tap the Notes tab to see a list of your notes and highlights. Tap any item to jump to that page.

Look up Words in the Dictionary

Long-press a word in the text to select it. A pop-up appears at the bottom of the screen with the definition. Tap on the pop-up to see the full definition. Tap elsewhere on the screen to close the pop-up and return to reading.

Kindle Books

Amazon Kindle books have their own proprietary format but there is a free Kindle Reading App for Android (http://bit.ly/1a2YJNe) so you can read Kindle books without owning a Kindle e-reader. With the Kindle app, you can not only read Kindle books, magazines, newspapers, and blogs, you can also shop in the Amazon Kindle store. This gives you access to more than two million paid Kindle titles and another two million free, public domain titles.

Registering Your Kindle App

In order to buy books and other content from the Kindle Store, you'll need to have an Amazon account and also register the Kindle App with Amazon. If you don't have an account, go to Amazon.com to create one. The first time you open the Kindle App you'll be prompted for your Amazon email address and password.

Amazon Cloud

Just as with Google Play books, the first time you open a Kindle book, it is downloaded to your device and you do not have to have a Wi-Fi or LTE connection to read it in the future, unless you delete it. All Kindle content that you purchase in the Amazon Kindle Store (even if the price is zero) is stored in the Amazon Cloud and you can retrieve it at any time.

Using the Kindle App

The Kindle app closely resembles the reading experience on the Amazon Fire tablets. The home screen has a Menu bar at the top of the screen. Tap the Menu icon at the left to see the main menu, which lets you view all items, items downloaded to your device, books, personal docs, newsstand items including magazines, newspapers and blogs, and go to the Kindle store. You can also access settings, app info, send feedback, and get help.

Other icons on the Menu bar give you access to the Kindle Store and let you synchronize the app. Synchronizing downloads any new content you've purchased as well as personal documents you've emailed to Amazon for conversion. See Personal Documents.

The Carousel occupies the middle of the screen and shows your most recently accessed content. Swipe left to scroll the Carousel, and tap any item to view it. If the item is not on the device, it will be downloaded from the Amazon Cloud. Items that have been downloaded are marked with a check icon. Long-press any Carousel item to see options for that item. The options vary for different content-for books, you can go to last page read, go to the beginning, download to Home or remove from device, and remove from the Carousel. Periodicals give you the option of marking issues to keep so they won't automatically be deleted after a period of time.

dirt examining them, knowing there was little I could do to prevent the blisters from going from bad to worse. I ran my finger delicately over them and then up to the black bruise the size of a silver dollar that bloomed on my ankle—not a PCT injury, but rather evidence of my pre-PCT idiocy.

It was because of this bruise that I'd opted not to call Paul when I'd been so lonely at that motel back in Mojave; this bruise at the center of the story I knew he'd hear hiding in my voice. How I'd intended to stay away from Joe in the two days I spent in Portland before catching my flight to LA, but hadn't. How I'd ended up shooting heroin with him in spite of the fact that I hadn't touched it since that time he'd come to visit me in Minneapolis six months before.

"My turn," I'd said urgently after watching him shoot up back in Portland. The PCT suddenly seeming so far in my future, though it was only forty-eight hours away.

"Give me your ankle," Joe had said when he couldn't find a vein in my arm.

I spent the day at Golden Oak Springs with my compass in hand, reading *Staying Found*. I found north, south, east, and west. I walked jubilantly without my pack down a jeep road that came up to the springs to see what I could see. It was spectacular to walk without my pack on, even in the state my feet were in, sore as my muscles were. I felt not only upright, but lifted, as if two elastic bands were attached to my shoulders from above. Each step was a leap, light as air.

When I reached an overlook, I stopped and gazed across the expanse. It was only more desert mountains, beautiful and austere, and more rows of white angular wind turbines in the distance. I returned to my camp, set up my stove, and attempted to make myself a hot meal, my first on the trail, but I couldn't

Reading a Play Book with the controls hidden. To see the controls, touch the center of the screen, and to hide them again, touch again. Reading with controls hidden is the normal mode in Play books as well as the Kindle and Kobo apps, and makes the reading experience as book-like as possible

The bottom of the Home screen is occupied by a list of Amazon-recommended books or the Kindle Select 25, which you can choose by tapping the buttons just above the list.

Reading Kindle Books

As an example of how the Kindle app works, here's how you read a book. You can open a book from the Carousel if you read it recently, or from the Menu if not. To see a complete list of your books, tap the Menu icon, then Books. Tap any book to open it. If the book is not on your device, it will download from the Amazon Cloud as long as you have a Wi-Fi or LTE connection.

Books open full screen by default. To turn pages, tap the left or right edges of the screen, or swipe left or right. Most books show the title at the top of the page and your reading progress at the bottom.

Tap in the middle of the screen to show the Menu bar at the top and the Progress bar at the bottom. To leave the book and return to the Books library, tap the Back icon at the upper left corner. Tap the Aa icon to change font size, margin size, line spacing, screen color, and brightness. (If automatic brightness has been set in the device settings, this setting will be disabled. To turn off automatic brightness, swipe down from the upper right corner of the screen to show the settings menu, then tap Brightness. Move the slider to disable automatic brightness. Now you can set brightness from the Kindle Aa pop-up.)

You can search the current book by tapping the Search icon and entering a search term with the virtual keyboard. Tap any item to jump to the page. Tap the back icon at the very bottom of the screen to return to the previous page.

While reading, you can tap the center of the screen and then the Menu icon at the upper right to access the Kindle Store, go to the table of contents and locations in the book, view your notes and marks, sync to the furthest page read, add a bookmark, and share your progress with others.

Bookmarks

To bookmark any page, tap the upper right corner. A blue bookmark appears. To remove a bookmark, tap the book mark.

To view your bookmarks, tap the center of the screen, then tap the Menu icon at the upper right, and select View your notes and marks. Tap any item to jump to that page. To return to your previous location, tap the back icon at the very bottom of the screen.

Personal Documents

You can email personal documents to the Send-to-Kindle email address associated with the Kindle app. Amazon will convert the documents and send them directly to the Kindle app when your device is connected to the Internet via Wi-Fi or LTE. Personal documents will show up in the Carousel after a few minutes, and also under Docs from the main menu.

Kindle app home screen

joined in, and we gave ourselves three months to use our best people and experiences from Viking, Pathfinder, and Polar Lander to look again at the whole Mars EDL concept from top to bottom. Perhaps by taking the lessons we had learned from the earlier projects, we could come up with ideas that might help. Brian's organization chart for our study team was drawn with a circle around each subtopic. There were so many circles that the chart looked like a tub full of soap bubbles, and we were soon calling ourselves "the bubbleheads." We dove in to address a wide range of EDL questions, among them these:

• Were there more reliable ways of ensuring that a lander could be guided to the desired point of entry at the top of the Martian atmosphere?

• What kind of information and imagery would we need from future Mars orbiter spacecraft to help us pick places on Mars that would be safe enough to land?

• Could we figure out how to guide our spacecraft to a smaller landing target area?

• How could we provide for slowing down a heavy spacecraft: Larger parachutes? Multiple parachutes? Larger rockets?

• How do we guide our landers so that they could land within an area the size of a city block instead of a large county? Could we keep track of where the lander was in relation to where we wanted it to land? Could the lander see where it was on its map, as an astronaut would? Or would we need to place a beacon on Mars for future landers to home in on?

• How might a lander be able to see big rocks and steep slopes, and avoid them?

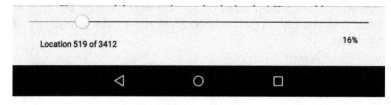

Reading a book in the Kindle app with the Menu and Progress bars shown

To see and edit your Kindle App address on Amazon.com, go to Your Account, then Manage Your Devices. (http://amzn.to/1cIwHXZ) Select your Android device from the list to see the assigned email address. To change it, click on Edit.

To send documents to the Kindle App, email them as attachments to the Send-to-Kindle email address. You can compress multiple documents in a ZIP file and send them all at once.

Amazon will only accept emails from approved email addresses. The email used to sign in to your Amazon account is approved by default, and you can add more email addresses on the Personal Documents Settings page (http://amzn.to/1b0zpqs).

Amazon can convert documents in the following formats:

- Microsoft Word (.DOC, .DOCX)

- HTML (.HTML, .HTM)

- RTF (.RTF)

- Text (.TXT)

- JPEG (.JPEG, .JPG)

- Kindle Format (.MOBI, .AZW)

- GIF (.GIF)

- PNG (.PNG)

- BMP (.BMP)

- PDF (.PDF)

Personal documents are archived in the Amazon Cloud for later retrieval, even if you delete the document from the Kindle App. You can disable this feature on the Personal Document Settings page. You may wish to do this if you send multiple revisions of the same document, to avoid cluttering up your Cloud storage.

Kobo Books

Like Amazon, Kobo (kobobooks.com) has a free app that is available in the Play Store (http://bit.ly/1EBCGf9), so you can read Kobo books on Android without having to own a Kobo e-book reader. The Kobo app is similar to the Kindle app, in that you can access the slide-out menu by swiping right from the left edge of the screen. From here you can access the Kobo Store and browse your library of books and magazines.

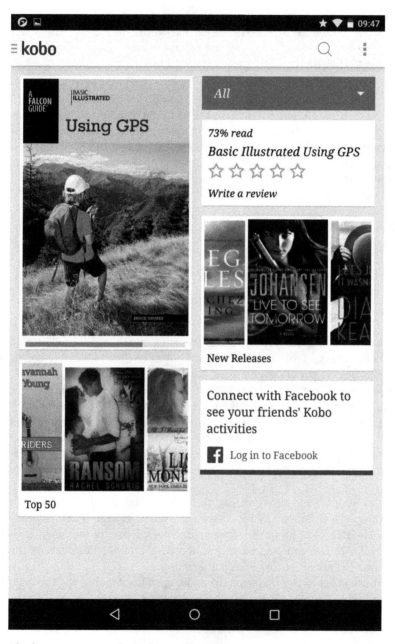

The home screen on the Kobo reading app

Free Books

Play Books

Free books are available from Play Books, usually as time-limited promotions. To see a list of free books, go to Play Books on the Nexus, then Shop, and scroll down until you see the Read for Free list of free books- it's usually the last category. Then tap See More.

Kindle Books

Many books are also offered for free in the Amazon Kindle Store, either as temporary promotions, or because they are in the public domain. To find free books, first search for or select a category of books on the Amazon Book Store on your computer. Then use the Sort By drop down at the upper right side of the screen to sort by Price: Low to High.

The easiest way to find free and 99-cent books is to use the tools at Kindle Nation Daily:

Free books: http://kindlenationdaily.com/kindle-nation-daily-free-and-bargain-book-listings/

99-Cent Books:

http://kindlenationdaily.com/knd-quality-99-centers-full-list-3/

Another great site for free and low-cost books is bookgorilla.com

Project Gutenberg

Gutenberg.org is the original free book site, predating even the rise of electronic book readers and tablets. Project Gutenberg now has more than 44,000 free books, mostly classics that are out of copyright and are now in the public domain.

The easiest way to get books from Project Gutenberg is to read it in the free Kindle App. There are several ways to do this, but these two are the easiest:

On your computer or your device, go to gutenberg.org, find a book, then download it in Kindle format. Then email the book to the Send-to-Kindle email address for the Kindle App on your Android device. When your Nexus is next connected to Wi-Fi or LTE, the book will be converted to Kindle format and sent to the device.

Or, on your device, go to gutenberg.org, find a book, and download the book in Kindle format. Now open a file manager such as ES File Explorer (http://bit.ly/HwRtQc) and go to the Downloads folder. Look for the new file- it will have a MOBI extension- and move it to the Kindle folder. For more on file transfer, see the chapter *Managing Files and Printing*.

After you use either method, to open the book, open the Kindle App. The book will show up in the Carousel and under All Items- but not under Books.

Library Books

You can borrow books from most public libraries using Overdrive, an app that comes with Android. Chances are your local library will let you borrow books with Overdrive. You'll need a library card and an online account on your library's website to borrow books.

To get started, open Overdrive, then tap the Overdrive menu icon at the upper left corner to open the menu. Then tap Add a library. You can either search for or browse libraries. Once you've located your local library, add it to Your Libraries by tapping the star next to the name.

Next, tap the Menu icon again and tap Bookshelf. Tap Add a Book. Tap Sign In at the upper right corner of the screen and enter your login and password for the library. Using the icons at the top of the library screen, you can browse by genre, check your Overdrive library account, get help, and search for books.

Once you find a book, tap on it to open the detail page. This page includes the book description, the loan period, the formats and number of copies available, and recommended books. To borrow the book, tap the Borrow button. You'll be given a choice of book formats to download. Choose EPUB to read in Overdrive, or Kindle to read in the Kindle App.

Assuming that you choose EPUB, you'll be asked to sign in with your Adobe ID, or will be asked to create one. After this one-time step is complete, the book will appear in your Bookshelf, which you can access from the main Overdrive menu.

Tap on a title to open it. Books open to the last page read. Tap the center of the screen to show the reading controls, which are similar to Play Books and the Kindle app.

To return a title, go to Bookshelf and long-press the title. Then tap Return. From this pop-up menu, you can also share the book via email, Facebook, Goodreads, and any other sharing app on your device.

Google Play Magazines

To shop for magazines, open Google Play and tap Magazines. As with books, you can browse by categories or search for a specific title. Once you find an interesting magazine, tap the magazine to view its detail page. You can buy single issues or subscribe to most magazines. Subscriptions come with a 14-day free trial.

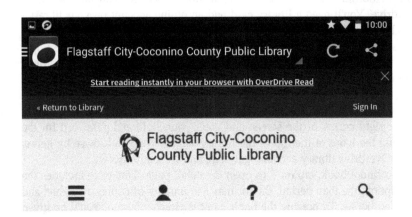

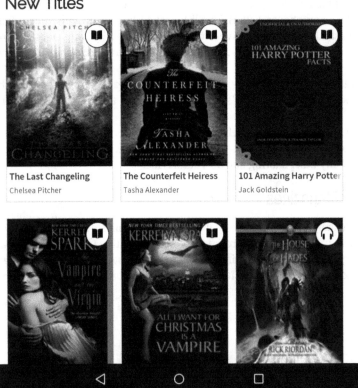

Browsing a local library for ebooks with Overdrive

Zinio Magazines

Another way to buy magazines is through the free Zinio app (http://bit.ly/1hKiZHy). Zinio features more than 5,500 magazines, which you can read in the Zinio app or at zinio.com in your computer's web browser.

The advantage of Zinio is that it's cross-platform. Not only can you read magazines bought at Zinio on your computer and on any Android phone or tablet, you can install Zinio on iOS devices including iPads and on Amazon Fire tablets. This avoids having to buy separate subscriptions for different platforms. Also, Zinio automatically archives back issues so you can retrieve them at any time, even after you've deleted them from your Zinio library. Another great feature of most Zinio magazines is Text mode. At the start of each article, you can tap Text at the bottom of the screen to view an article without images or ads.

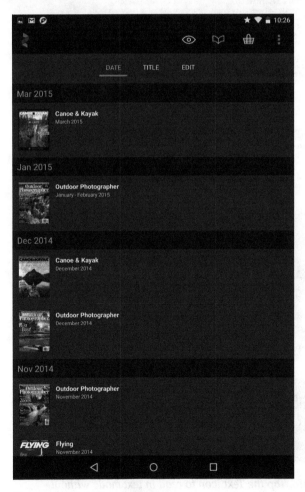

Your magazine library on Zinio

Last summer, brother and sister Sawyer Kesselheim, 21, and Ruby Zitzer, 19, completed a 1,000-mile crossing of the Canadian Barren Lands with their cousin, Quinn Mawhinney, 24, and friend, Kelly Kjorlien, 22. The expedition from Great Slave Lake to Hudson Bay via the Hanbury and Thelon rivers spanned 41 days amidst the worst wildfire season in Northwest Territories' history.

FIRE AND ICE

Everything but locusts in the Canadian Arctic

BY CONOR MIHELL

RUBY: I started thinking about the trip in my junior year of high school. I wanted something long—at least 30 days. Starting in Yellowknife meant we could drive in and fly out. I wanted to be on the tundra and go over the height of land and paddle a mix of lakes and rivers.

SAWYER: It was originally going to be an all-girls trip but Ruby needed one more person. It was an opportunity that does not come around often. My sister and I work well together and I've always wanted to plan a trip like this with her. I was excited about being out there without distractions and enjoying the challenges and simplicity of camping and traveling.

RUBY: On Day Eight we were stopped by ice on Great Slave Lake. For a few days we could weave and shift through the ice but then it became more solid. We were making two miles per day. It was exhausting and awful. It was early in the trip but we were already worried about running out of food.

SAWYER: When we saw the smoke of wildfires, we expected exploding trees and raging fires like at home in Montana. But up there it was more of a smoldering ground fire because the trees are so small. The smoke was the biggest challenge because it made navigation tough. There were days we were unsure if we could move due to low visibility.

RUBY: Once we got through the ice we had to push really hard—getting up at 4 or 5 a.m. and paddling 30- to 40-mile days. We had 15 straight days of headwinds. That meant few bugs, but I'd take mosquitoes over wind.

KELLY: The emptiness and expansiveness of the tundra draws you in. It's a playground of sorts. Since we needed to spend so much time on the water, it was always a nice break to get out of the boats and romp around in the pillows of moss.

RUBY: We grew up doing big trips with our parents. There were definitely times when Sawyer and I wondered, 'What would our parents do?' We about the trip when we got home.

SAWYER: Ruby is so easy to travel with. She's also one of the toughest people I've ever met. She was always first in line to carry the canoes.

RUBY: Oh yeah, I want to go back. The tundra is one of the most powerful places I know.

74 | canoeroots.com

PHOTO: RUBY ZITZER

TEXT

Reading a magazine in Zinio. Tap the Text icon to read in text mode without graphics or ads

Keeping Up with the News

All the major news sources such as CNN, NBC, NPR, the New York Times, AP, Reuters, and many more have dedicated free apps. Some of these apps give you unlimited access, while others such as the New York Times require a digital or print subscription for unlimited access. To find news apps, go to Google Play on your device or a computer, select Apps, and search for "news" (http://bit.ly/1iKkm71).

Google News

Another way to keep up with the news is a news feed such as Google News (news.google.com). Google News aggregates news from primary sources worldwide, so you can see how the same story is reported in different countries, not just your home country. And you can create custom filters to show news topics of your choice.

You can access Google News via a web browser on Android or a computer, or choose from many apps (http://bit.ly/1a5cVqb).

Flipboard

Flipboard (http://bit.ly/14VarZ7) is one of the best news apps. Flipboard can display feeds from social sites such as Facebook, LinkedIn, Google Plus, and Twitter, as well as Google News and many others. Flipboard can replace dozens of separate news and social apps. You can choose from a list of social and news media accounts and add news feeds by category.

Flipboard then presents your news and feeds in a beautiful magazine layout that you flip through in a natural manner that is a perfect match for touchscreen devices. You can tap on any story to open the full story. From there, you can share the story via email and social sites as well as open the original web page.

Feedly

Feedly (http://bit.ly/16DaObq) is a news aggregator app that is especially good for blogs and RSS feeds, though you can certainly use it as a news feed.

RSS (Rich Site Summary- sometimes called Really Simple Syndication) is a feature found on many websites which lets you subscribe to automatic updates. These stories and updates are pushed to Feedly so you can keep up with websites without having to go to each individual site.

Look for the RSS icon,

on a website of interest. For example, on my website grandcanyonguide.net, the RSS icon is in the upper right corner of the home page. However, many websites have RSS feeds but don't display the icon. To add a site, open Feedly and tap the Menu icon at the upper left corner. Then tap Add Content. Type the address of the website. It will appear in search results if a feed is available. Tap the + to add the site, then choose the category.

To remove a feed, scroll down the list of feeds to Settings. Tap Settings to expand it, then tap Edit Content. Tap the check box next to all the feeds you wish to remove, then tap Remove.

From the Add Content button, you can also search for a news feed or choose from many different sources. Once you find a source you like, just tap the + icon at the top of the screen to add it to your feeds. You can organize your feeds into categories that you create, such as World News, Blogs, etc.

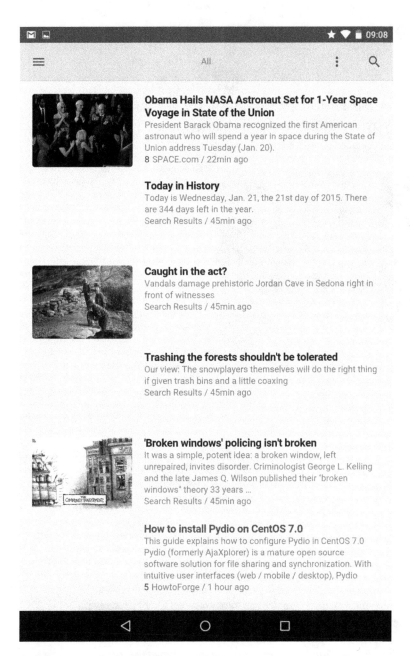

Article summaries in Feedly

Obama Hails NASA Astronaut Set for 1-Year Space Voyage in State of the Union

8 SPACE.com / 22min ago

President Barack Obama recognized the first American astronaut who will spend a year in space during the State of Union address Tuesday (Jan. 20).

| Tweet | Visit website |

Story detail in Feedly. Tap Visit website to see the complete story

Watching the Weather

All right, I'd better admit right up front that I'm a weather nut. As a lifelong hiker and outdoor type, the weather has always affected what I like to do. After I became a professional pilot, weather started affecting my work too. So I have good excuses. Fortunately for me and all you other weather nuts out there, the Internet and mobile devices have put almost unlimited weather information literally at our fingertips. I can only hit the high points here, and I'm sure I've missed some cool apps (pun intended.)

Weather Apps and Widgets

There are many ways to arrange weather icons and widgets to customize your device. This chapter focuses on the weather apps and widgets themselves. For more on setting up your Home screen and Favorites area, see *Customizing Your Device*.

Personally, as a U.S. resident, I prefer to get my weather information directly from the official source, the National Weather Service (NWS). All of the private weather services and mass media get most of their weather information from NWS, so why not eliminate the middleman? On the other hand, the private services sometimes present weather information in a more useful way.

Weatherbug

Weatherbug (http://bit.ly/1cRGIC7) is a great choice for detailed international weather. It has a very attractive and useful home screen, which by default automatically displays today's weather for your current location. Tiles show today's forecast, radar, lightning data weather cams, weather photos, and the pollen forecast. Tabs at the top of the screen allow you to display the 10-day and hourly forecasts, as well as detailed climate and astronomical information. A window at the bottom of the screen shows the top weather news from around the country.

Tapping the Menu icon at the upper left corner gives you access to much more weather information, including radar, severe weather alerts, lightning maps, hurricane information, media reports, weather cams, video reports, and forecasts tailored to specific activities.

The weather map can be customized with many overlays, including U.S. radar, Canadian radar, worldwide radar, U.S. Satellite, worldwide satellite, temperature, wind speed, humidity, and U.S. high and low temperature forecasts. The Google Maps base map can show traffic and satellite views, and the transparency of the overlays can be varied.

Settings allow you to customize Weatherbug to suit your needs. You can change the background color and image, change the order of the tiles on the home screen, set preferences for units, location, alerts, and check FAQ's and get help. You can also show current weather in the Notifications bar.

Local forecast in Weatherbug

Weatherbug comes with two small widgets that you can place on your home screen to show the current weather and tap for quick access to Weatherbug. Weatherbug Elite is the paid, ad-free version.

Wunderground

Wunderground.com (Weather Underground) is famous for its Wundermap which can display different weather layers for the entire planet except for the polar regions. The map defaults to your current location, and you can save as many locations as you want. Tapping the current location bar at the top of the map shows you current, forecast, and hourly weather for that location.

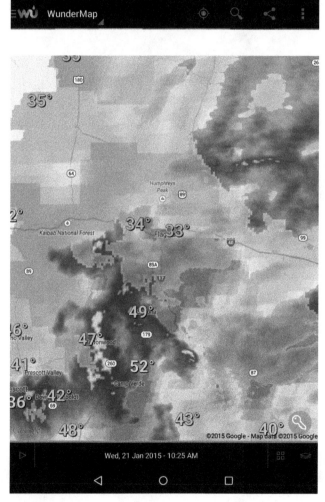

Wundermap

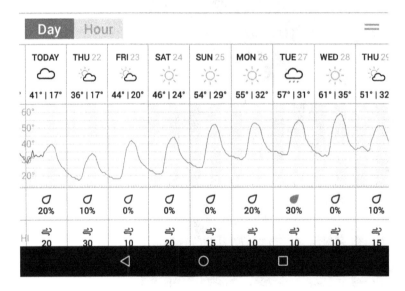

Local weather forecast in Wunderground

Map overlays include radar, satellite, wildfire risk and activity, hurricane activity, severe weather, fronts, and web cams. The map background can be set to show radar, weather station temperature, precipitation, or wind, and regional temperature tints. When weather stations are shown, you can tap on any station for the current, forecast, and hourly weather.

Wunderground's app (http://bit.ly/1u7BRKe) has a widget which can be placed on your home screen to show current weather and the forecast and give direct access to the Wundermap.

Since it's the original weather site on the web it's not surprising that the Wunderground app has so much power.

The Weather Channel

As the only 24-hour cable TV weather source, it's not surprising that The Weather Channel focuses on weather videos from around the world. Their Android app (http://bit.ly/17MHLTY) has simple screens that can be tapped to show details. It defaults to the current weather at your present location. Tapping the large weather icon opens the weather and astronomical details for the day. You can touch the tabs at the top of the screen or swipe sideways to show radar, the current weather, the forecast, and weather videos.

NOAA Weather+

NOAA Weather+ (http://bit.ly/1xKqWCd) is a paid app that presents NWS daily and hourly forecasts, as well as radar images. From the app, you can directly access satellite images as well as the technical forecast discussion and the NWS web site.

You can tap the NOAA Weather+ icon at the top left of the screen toggle between the map view and a list of saved locations. You can also search for and save other locations by tapping the search icon at the top of the screen. Then you can jump to any saved location by choosing it from the list. Long-press any location name to delete or rename it.

At first glance, NOAA Weather+ doesn't seem to offer as much weather information as some other apps. However, since it offers direct access to the NWS web site, you can get any weather product available from NWS, including digital forecasts, climate data, severe weather and hurricane information, satellite images, and much, much more.

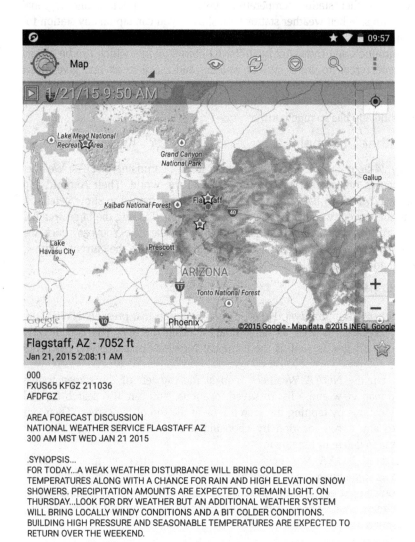

Flagstaff, AZ - 7052 ft
Jan 21, 2015 2:08:11 AM

000
FXUS65 KFGZ 211036
AFDFGZ

AREA FORECAST DISCUSSION
NATIONAL WEATHER SERVICE FLAGSTAFF AZ
300 AM MST WED JAN 21 2015

.SYNOPSIS...
FOR TODAY...A WEAK WEATHER DISTURBANCE WILL BRING COLDER
TEMPERATURES ALONG WITH A CHANCE FOR RAIN AND HIGH ELEVATION SNOW
SHOWERS. PRECIPITATION AMOUNTS ARE EXPECTED TO REMAIN LIGHT. ON
THURSDAY...LOOK FOR DRY WEATHER BUT AN ADDITIONAL WEATHER SYSTEM
WILL BRING LOCALLY WINDY CONDITIONS AND A BIT COLDER CONDITIONS.
BUILDING HIGH PRESSURE AND SEASONABLE TEMPERATURES ARE EXPECTED TO
RETURN OVER THE WEEKEND.

&&

.DISCUSSION...

A radar map and the technical forecast discussion in NOAA Weather Plus

Finding Your Way with Maps and GPS

Since most Android devices have a GPS (Global Positioning System) receiver, they can be used as a powerful mapping and navigation device. Using Google Search and Google Voice, you can find places and get directions right from the search bar on the Home screen. These maps and directions are powered by Google Maps, but you can install other navigation and map apps to extend the capability of your device.

There are several different types of maps that are useful on Android devices. The most popular are street maps for city and highway navigation, and topographic (topo) maps for backcountry and outdoor navigation. Most everyone finds street mapping useful, while topo maps are primarily used by hikers, hunters, and others engaged in outdoor activities. Aviation and marine charts are specialized maps used by pilots and boaters.

Warning!

Never depend solely on an electronic device for wilderness navigation. Even though many Android phones and tablets have remarkable battery life, it can't be compared to a paper map with a "battery life" of infinity. Also, many mapping apps require an Internet connection for access to maps. You won't find Wi-Fi hotspots in the wilderness, and LTE is not available in most remote areas.

Dedicated trail GPS receivers are still far superior to tablets and phones for backcountry use. Trail GPS receivers have more sensitive receivers that can maintain a satellite lock under more difficult conditions. They also have longer battery life, from 12 to 24 hours or more, and most use field-replaceable AA or AAA batteries. Bottom line- always carry paper maps and a reliable compass to back up your electronic navigation, whether tablet, smart phone, or trail GPS receiver.

Some covers and cases contain magnets, which will interfere with the electronic compass in your device and cause mapping apps to work incorrectly and display incorrect directions. Some apps warn you, while others don't.

Street Maps

Street and satellite maps let you find places and navigate nearly anywhere in the world. However, mapping an entire planet is a big job, even with modern tools. Errors can and do creep in, so back up your electronic navigation with paper maps and guidebooks. And use common sense- don't be one of those drivers who blindly follow GPS instructions right off a closed bridge into a river!

Google Maps

Google Maps is the 500-pound gorilla of electronic mapping, and since your phone or tablet is a Google device running a Google operating system, Android, it's no surprise that Google maps are tightly integrated with the device. You can search for places and get directions using the Google Search bar that is installed on the home screen by default.

Google Search

There are two ways to search for places (or anything else) with the Google Search bar. You can tap the bar and type in a search term with the virtual keyboard, or you can tap the microphone icon at the right end of the search bar. You can also say "OK, Google" to activate voice search. When prompted, speak your search term. Voice search works well with common words, but you may have better luck with the keyboard with odd place names.

For example, to locate the nearest coffee bar, just tap the mic icon and say "coffee bar". After a few seconds, Google Search responds with a voice prompt and a list of nearby coffee bars.

If you want to search for a specific place, just say the name. Saying "Starbucks" locates all the nearby Starbucks locations. The voice prompt varies with the search results.

The results list may include websites as well as locations. You can tap any of the websites to open that site in a web browser.

Locations will be shown on a Google Map. You can tap the map to view it full screen, or scroll down to see the locations. Tap the web address or title of the listing to go to that location's website. Tap More Info for a detailed map and information such as hours of business. Tap Directions or the location marker for directions, then tap Start to start GPS navigation.

Location Services

In order to calculate directions to your destination, your device has to know where you are. It uses a feature called "Location Services" to find your location, using Wi-Fi networks and GPS satellites. If you attempt to navigate when Location Services is turned off, you'll be asked if you want to turn it on, and taken to Location settings if you answer yes.

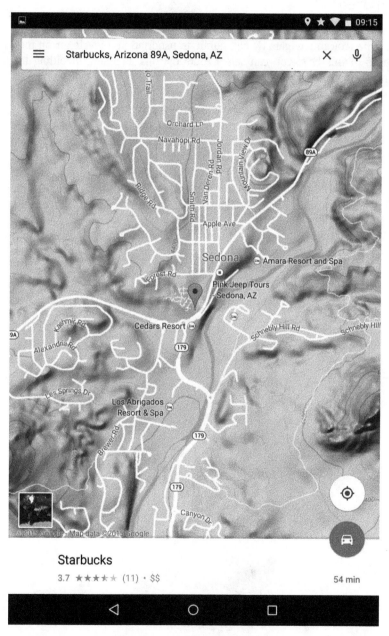

Google Maps search results

Unless you are trying to maximize battery life, you should have both GPS and Wi-Fi location turned on. GPS is more accurate, but Wi-Fi works in places such as buildings where the device can't receive the GPS satellites.

Navigation Options

Tap the From and To location bar at the top of the screen to show navigation options. Tap the icon to navigate by car, public transportation, bicycle, or on foot. At the bottom of this box, you can select route options, including avoiding toll roads and highways.

Another box shows the distance and time under current traffic conditions. Below that, you can view alternate routes to your destination.

When you tap Start, your device starts voice-prompted turn-by-turn navigation to your destination. The device uses the GPS satellites to track your location so the tablet must have a clear view of the sky. Usually, a front seat passenger holding the device in front of them gives the GPS receiver a good enough view of the sky. If you plan to use your phone or tablet extensively for vehicle navigation, you might want to install a vehicle mount and power cord. See the chapter *Finding Accessories*.

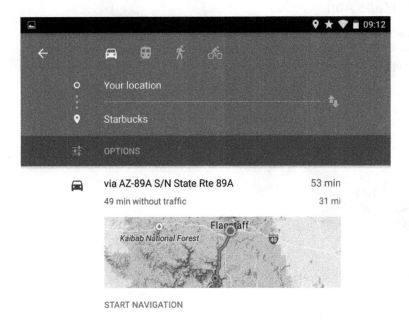

Navigation options in Google maps include driving, public transit, walking, and cycling, selected by tapping the icons at the top of the screen

You can stop navigation by tapping the X at the lower left corner of the screen. That takes you back to map view.

Google Maps App

The full screen Google map that you used in the example above can be opened directly via the Google Maps app, which is pre-installed on Android. No matter how you open it, Google Maps has many of the features of the web-based desktop version.

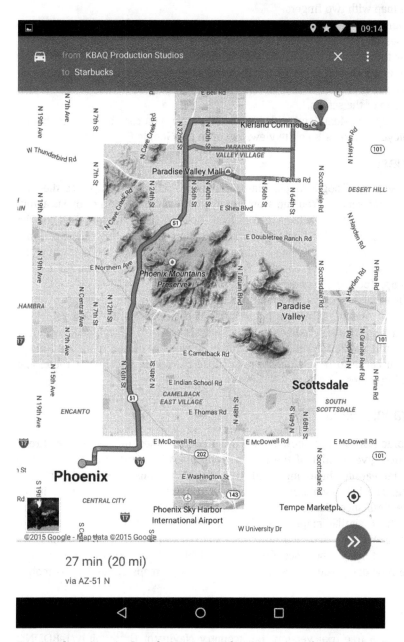

Driving directions in Google Maps. Tap the start icon to begin navigating

Running Android

You can center the map on your location by tapping the Center icon at the lower right corner of the map. In order to do this, Location Services must be on, as described above. You can swipe to move the map, pinch-zoom in and out, and rotate the map with two fingers.

You can search for places by tapping the search bar at the top of the screen. You can search for categories such as restaurants, hotels, gas stations, or airports, search for the name of a business, a geographic name, or a street address. When you tap the search bar, a list of recent searches pops up, as well as your search history. To see more of this list, tap the Hide Keyboard icon left of the Home icon at the bottom of the screen.

By tapping the Menu icon at the upper left corner, you can show traffic, public transit, bicycling routes, satellite view, and open the Google Earth app.

MapQuest

An alternative to Google Maps, MapQuest (http://bit.ly/16HkcLq) was the first web-based map service. You can search by typing in a search term or by speaking. Icons along the bottom of the screen let you show points of interest (POI) on the map, including hotels, food, gas, airports, shopping centers, coffee shops, parking, grocery stores, schools, hospitals, and movie theaters. Icons on the map show the location of the POI you've enabled. If there are no matching POI on the map, a box pops up. Zoom the map out by pinch-zooming.

MapQuest also shows traffic and has a satellite view. The street map can be shown in day and night modes. Night mode is easier to see in a dim vehicle at night. You also manually edit your route and set roads to be avoided, including highways, toll roads, ferries, unpaved roads, seasonally closed roads, and direct routes. You can also have MapQuest calculate the shortest or fastest route, and also walking routes.

Topographic Maps

Topographic (topo) maps show terrain using contour lines, unlike street and road maps which give no hint of the terrain. Topo maps have many uses aside from outdoor recreation, including land use planning, construction, zoning, and surveying. The U.S. Geologic Survey is the primary source of topo maps in the U.S. The Ordnance Survey produces topo maps for Great Britain, and other countries have similar mapping services. Commercial maps are produced using government mapping data and other data sources to create specialized maps for recreating and many other uses. Formerly available only as expensive paper maps, there are now many sources of free or low cost topo maps online and as mobile apps.

Backcountry Navigator Pro

Available in a free trial version, Backcountry Navigator (http://bit.ly/1a8DDNs) can display topo and other specialized maps for many countries. Map selections

110

can be downloaded to your device for use offline, an especially handy feature for backcountry use. To download maps, go to the map and area of interest, then tap the Map Layers icon. Tap Select Areas for download, then draw rectangles on the screen to select the areas to download. You can select multiple areas. When finished, tap Download, and then from the pop-up, Begin Download. You'll be shown how much space the downloaded maps will require, and the amount of free space on the device. Tap Proceed to start the download.

Backcountry Navigator makes use of GPS for location and navigation, and you can create, download, and upload waypoints and tracks.

Backcountry Navigator comes with CalTopo 1:24,000 topo maps of the U.S., which are based on U.S. Geological Survey maps; OpenCycle map, which covers many countries worldwide; USGS topo maps; and U.S. Forest Service topo maps.

If you upgrade to the Pro version, you can choose from a huge selection of paid maps, such as Bureau of Land Management boundary maps showing private and public land, hunting unit maps, lake contour maps, and trail maps.

Other Maps

Google Earth

The Google Earth app (http://bit.ly/1eox7GN) is pre-installed on Android. It gives you access to the worldwide satellite and street view imagery that has made Google Earth such a useful tool. You can also access it from the menu in Google Maps.

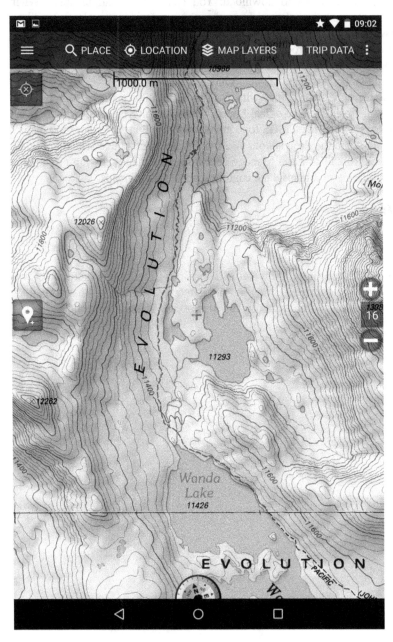

Detailed topographic map in Backcountry Navigator Pro

Navigation in the Android app is even easier than in the desktop app. Swipe any direction to pan the image, and rotate the image with two fingers. Pinch zooming works as expected to zoom in and out. The only action that is not obvious is tilting the view- do this by placing two fingers side by side and moving them up or down.

Tapping the N icon reorients the image to north up and removes any tilt, and tapping the Center icon centers the image on your present location. From the Menu icon you can share the image, access settings, leave feedback, get help, and view a brief tutorial. You can search for a place by tapping the search icon at the top of the screen.

Tap the Menu icon at the upper left corner to slide out the Layers and Map selections.

Sky Map

Looking in the other direction, Google Sky Map (http://bit.ly/HC9dK1) puts a planetarium on your phone or tablet. When first opened, Sky Map uses the tablet's sensor to automatically show the part of the sky you're looking at as you hold the tablet up. You can switch to manual mode by tapping the screen to bring up the control icons, then tapping the mode toggle icon at the lower left corner.

In manual mode, drag to pan the map and pinch-zoom to move in and out. You can rotate the view with two fingers.

Icons on the left side of the screen let you toggle various sky objects on and off. From top to bottom they are- stars, constellations, Messier objects, planets, and meteor showers. The last two icons toggle the right ascension/declination grid and the horizon line and nadir/zenith markers.

From the main menu icon at the lower right corner of the screen, you can search for objects by name, toggle the screen between day and night mode, view a gallery of sky objects, change the date and time of the sky view, and change settings.

Settings includes sky object toggles, which are the same as the pop-up screen controls, and change location settings, sensor settings, sound, and enable multi-touch rotation. You can also get a brief help screen.

Aviation Charts and Flight Planning

There are many free and paid aviation apps (http://bit.ly/1gKuv76)- far too many to cover here. I'll just describe a few that I've found to be helpful as an airline passenger and as a charter and air tour pilot flying small aircraft.

FlightAware

FlightAware (http://bit.ly/1eMUsSM) is a free aircraft tracking app that can display information on a moving map for scheduled airline flights worldwide, as well as most other aircraft on an instrument flight plan, which includes most jet and turboprop aircraft (it doesn't track aircraft on visual flight plans.) Since

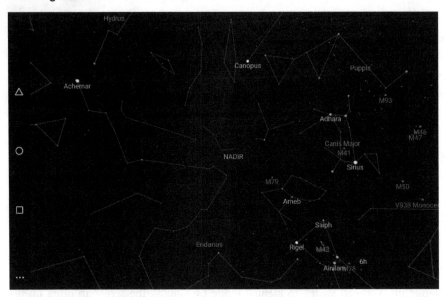

Google Sky Map

FlightAware gets its data from air traffic control, it is far more accurate than airline web sites or airport flight status boards.

Avare

Avare (http://bit.ly/1aTM9TA) is a free aviation flight planning and charting app that allows you to download maps for use offline. As the app warns you, it is not an FAA-certified GPS or navigation system and should be used to supplement, not replace, certified aircraft equipment and official charts. Having said that, Avare is extremely useful, both for preflight planning and enroute navigation. In the air, it works as a GPS moving map at a cost hundreds of dollars cheaper than dedicated electronic flight bag (EFB) devices.

Aviation Weather

Aviation Weather (http://bit.ly/HU5XtX) is a free app that gives you direct access to National Weather Services aviation weather information. Though it is not a substitute for an official, legal weather briefing, it's an excellent flight planning tool.

Note that both official FAA online weather briefing sources are accessible from your phone or tablet. CSC DUATS (duats.com) doesn't have an app but does have a clean and simple mobile web site. DTC DUATS (duat.com), has a free Android app (http://bit.ly/1bv0Cnl).

Marine Charts and Navigation

Not being an active user of marine charts and navigation, I can only point out that there are many apps (http://bit.ly/18sH52E).

Marine Navigator

Marine Navigator (http://bit.ly/17uGCSt) is an example. It gives you access to marine charts worldwide, and the paid version lets you easily download charts for use offline. On the water, it works as a GPS moving map navigator.

GPS

Since most Android devices have GPS, many apps take advantage of the precise positioning that is available. Of course, GPS only works when the tablet has a good view of the GPS satellites, and may not work among tall buildings, indoors, and outdoors in places such as narrow canyons.

That said, the GPS sensor is a valuable feature and greatly expands its utility. In addition to the mapping and navigation apps I've already described, you can get specialized GPS apps that display more GPS information and allow you to work with waypoints, tracks, and routes directly. There are also many GPS based apps for specialized activities, such as photography (see the section *Photographer's Aids*) and geocaching.

GPS Essentials

GPS Essentials (http://bit.ly/1iPb3mg) turns your device into a full-function GPS receiver. There are over a dozen main screens, many of which have a customizable dashboard area where you can place GPS values chosen from a long list.

The main screens include a full-screen dashboard and a geo-referenced camera app that records your GPS position and has a HUD (heads-up display) which shows azimuth, tilt, and a horizon line. A compass page shows your heading and the bearing to your destination, if one has been set.

You can choose from a number of maps to display, including MapQuest and Google Maps. You can display, create, and edit waypoints, routes, and tracks. You can also share GPS data with others. A satellite status page shows which GPS satellites are in your sky and which are being used to compute your location. You can also view the geo-referenced photos you've taken with GPS Essentials.

The app supports most map datums and coordinate systems which means it can be used anywhere in the world. In addition, GPS data can be imported and exported in KML and GPX format so you can work with GPS data in Google Earth, other mapping programs and apps, and with most dedicated handheld GPS receivers.

c:geo

c:geo (http://bit.ly/1hPGWNX) is for geocaching- the sport of finding, logging, and placing small hidden caches. Geocaching has exploded worldwide and there are millions of caches. For more information on the sport, check these websites: geocaching.com and opencaching.com. c:geo uses geocache data from geocaching.com, the original and still the largest geocaching website.

c:geo displays a live map of nearby geocaches, automatically centered on your location. You can pan and zoom the map, and select from Google map and satellite views, as well as open source maps such as OSM Mapnik and Cyclemap. You can save maps and caches for viewing offline, an especially valuable feature for owners of Wi-Fi only tablets.

Other screens let you see a list of nearby caches, stored caches, and search for caches by coordinates, address, geo code (each geocache in the world has a unique 6-character code name), keyword, user name, and owner name. You can also set a conventional destination and navigate to it. And you can filter caches shown by logged and cache type.

Once you've chosen a cache, you can use c:geo to navigate to it and log your visit if you find it.

Don't forget to store cache and map data on your device before heading out on the hunt, because you won't be connected to the map and geocache servers once you're out of range of Wi-Fi. Even owners of LTE tablets and phones may find themselves out of range in rural areas.

Another way to use c:geo is to upload geocache data to a dedicated handheld GPS receiver. You can export single geocaches or lists of geocaches as GPX files and send them to file sharing services such as Dropbox or email them as attachments. Then connect your GPS receiver to your computer and transfer the GPX file to the device.

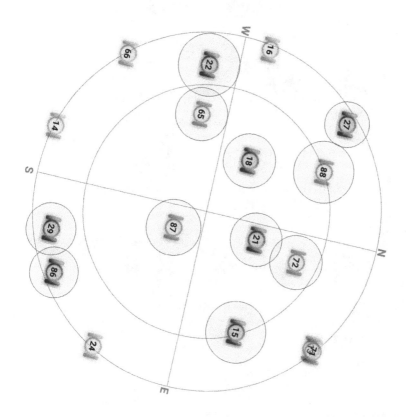

Satellite status screen in GPS Essentials

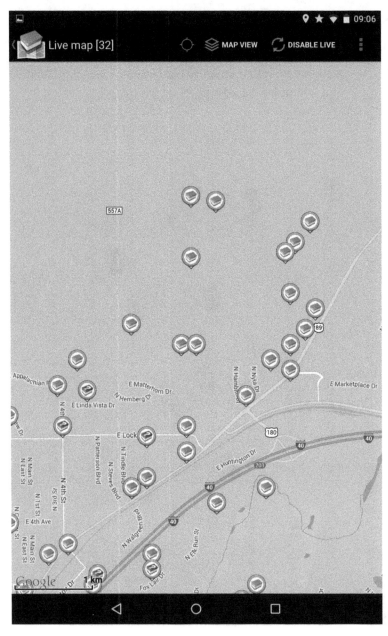

Live map of geocaches in c:geo

Your Pocket Office

While an Android tablet or phone can't do the heavy lifting of a notebook or desktop computer, you can still accomplish quite a bit of office-related work on one.

My own work is a good example. As an outdoor writer and photographer, my primary computer work involves creating and editing book files, editing and storing photos, and creating vector graphics and illustrations. For all of these tasks I can work most efficiently on my desktop system, which has two monitors, a fast processor and video card, a massive amount of disk space, and a full size keyboard. For photo editing and graphics work I use a precision graphics tablet. I also have a wired Ethernet connection to high speed Internet and a network-attached file server for backup. And of course I have printers- a color laser and a photographic-quality inkjet.

I can also do this work on my laptop, at a slight cost in efficiency due to the single screen and notebook style keyboard. But netbooks and tablets just don't cut it- my netbook doesn't have the processor power to edit photos or run Adobe Illustrator. And I haven't found a tablet that can handle book length documents, high end photo editing, and vector graphics.

That said, I can still accomplish a lot of work on Android. The nature of my writing requires a lot of Internet research and reference to maps, both of which are easily done on a fast tablet. And of course there's emails to and from editors, website maintenance, blogging, note taking, and managing the ebooks and print on demand books I've published under my own imprint. While the cameras on my phone and my tablet don't produce images of high enough resolution or quality for publication, I can still use them for location scouting and taking scouting or record photos.

I can also work with Microsoft Office and LibreOffice documents, including word processing, spreadsheet, and presentation files. There are several apps that support such documents. Even though they can't handle book-length files, I can still take notes, outline books, work with sales data downloaded from Kindle Desktop Publishing and CreateSpace, and work on presentations for book signings and other events.

For such text and number intensive work, you may find a physical keyboard and mouse useful. Fortunately, you can connect a Bluetooth or USB keyboard and mouse to Android devices. See the section Keyboards and Mice for details.

Working with Office Documents

There are many office apps which can work with common office file formats, such as word processing documents, text files, spreadsheets, and presentations. Here are just a couple that I've found useful.

Kingsoft Office

Kingsoft Office (http://bit.ly/199nYzz) is a good app for working with Microsoft word processing, spreadsheet, and presentation documents. You can display PDF documents as well.

AndrOpen

AndrOpen (http://bit.ly/18gYKfA) is a port of LibreOffice/OpenOffice for Android. It lets you work with LibreOffice formats for word processing, spreadsheets, presentation, drawing, math, and databases.

In my experience, both of these these apps are great for working with small documents, but book-length files make the apps sluggish and unusable.

PDF Documents

There are plenty of Android apps that display PDF documents, but the best still seems to be Adobe's own Adobe Reader app (http://bit.ly/188ZDf5). This makes sense, because after all Adobe invented the PDF (Portable Document Format) file format.

Text Notes

As with text-editing and word processing apps, there are many note-taking apps. These are two that I've used and find useful.

ColorNote

For keeping notes in a classic sticky note format, try ColorNote (http://bit.ly/17f8iLc). This app mimics sticky notes with a widget that shows your latest note. You can set the note color, archive notes, send and share notes, and set reminders.

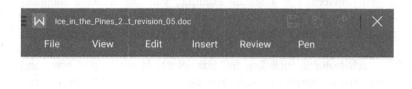

Ice in the Pines will be conducted in accordance with the rules and regulations of U.S. Figure Skating, as set forth in the current rulebook, as well as any pertinent updates which have been posted on the U.S. Figure Skating website.

This competition is open to all eligible, restricted, reinstated or readmitted persons as defined by the Eligibility Rules, and is a currently registered member of a U.S. Figure Skating member club, a collegiate club or an individual member in accordance with the current rulebook. Please refer to the current rulebook for non-U.S. Citizens.

TEST SESSION:
As time permits, a test session will be held on Saturday evening. Skaters desiring to test should submit an Arizona Interclub Test Application with a check for the test fees to FFSC, PO Box 3663, Flagstaff, AZ 86003 by August 9, 2014. Application forms are available from www.FlagstaffFigureSkatingClub.com/content/testing. Test fees can also be paid via PayPal on line.

ELIGIBILITY RULES FOR BASIC SKILLS PARTICIPANTS:

The competition is open to ALL skaters who are current eligible (ER 1.00) members of either the Basic Skills Program and/or are full members of U.S. Figure Skating. To be eligible, skaters must have submitted a membership application or be a member in good standing. Members of other organizations are eligible to compete but must be registered with the host Basic Skills Program/club or any other Basic Skills Program/club.

Eligibility will be based on skill level as of closing date of entries. All Snowplow Sam and Basic Skills 1-8 skaters must skate at highest level passed or one level higher and NO official U.S. Figure Skating tests may have been passed including MIF or individual dances.

For the Free skate 1-6, Test Track and Well Balanced levels, eligibility will be based only upon highest free skate test level passed (moves in the field test level will not determine skater's competitive level). Skaters may skate at highest level passed OR one level higher BUT not both levels in the same event during the same competition.

It is very important to the success of the competition that skaters are placed in the correct divisions. If, for whatever reason, the Local Organizing Committee discovers that a skater has been placed in a category that is below their class level, the chairman and referee will theoption to move the skater into the proper division, even if this has to be done the day of the competition. This will ensure that every event is as fair as possible to the competitors. Please be sure to check for the director/instructor's signature confirming the level of the skater.

ELIGIBILITY/TEST LEVEL FOR NON-QUALITFYING EVENTS:

Test level: Competition level is the highest test passed as of the entry deadline in the discipline the skater is entering. Entrants may skate one level above that for which they qualify, but they may not skate down in any event. Skaters who placed in the top four in a final round of their last qualifying competition in their divisions must move up one level, except for novice and higher.

This event is a standard U.S. Figure Skating Nonqualifying Competition

Editing a file in Kingsoft Office

EverNote

Probably the ultimate note-taking system for tablets, phones, and computers, EverNote (http://bit.ly/18ak6ZW) uses Cloud storage to synchronize your notes between all your devices. For a simple example, you can create a shopping checklist on your computer and it will automatically synchronize to your Nexus. While shopping, you can check off items. Next time you look at your shopping list on the computer, the checked-off items will be updated.

Powerful search features, including full text search, help you find notes and information that you've forgotten about. You can also tag notes to make them easier to find.

You can take photos of white boards, presentations, receipts, and any other paper document and store them in EverNote. Class and meeting notes, draft agendas, and any other on-site notes can be tagged and labeled for easy retrieval. You can also password protect notes to keep confidential information such as password lists secure.

Voice Notes

Smart Voice Recorder

Smart Voice Recorder (http://bit.ly/1hmAv7g) turns your phone or tablet into a sophisticated digital recorder for taking voice notes. All the expected controls for recording, pausing, and playback are present, as well as settings for skipping silence and adjusting microphone sensitivity. Recordings can be shared via email and social networks, and can be sideloaded to your computer.

Doing Research

Aside from the access to the Internet you have through Chrome and other web browsers, there are also specialized apps which can make research a lot easier.

Wikipedia

Wikipedia, the online, user-written encyclopedia that is rapidly becoming an important research resource, has its own free app (http://bit.ly/1hQtfhG). This allows you to go directly to the encyclopedia and enter a search term. Of course, you can still use a web browser, which lets you open the desktop site rather than the default mobile site.

Pocket

Pocket (http://bit.ly/1c3i3JE), formerly Read It Later, lets you clip a web article, image, or video and save it for later viewing offline. You can log into Pocket (getpocket.com) on your desktop browser and read or watch your saved items, or open the Pocket app on the Nexus and see your saved items. Pocket formats text into a very pleasing and easy to read magazine format and gets rid of extraneous graphics and ads.

To save to Pocket from apps that are sharing enabled on your device, tap the Sharing icon or open the main menu and look for a Sharing choice. Then select "Save to Pocket."

To save to Pocket directly from desktop browsers, you'll need to install the Pocket extension. All popular browsers have Pocket extensions. Many websites, such as new and blog aggregation sites like feedly.com, allow you to save to Pocket. Just look for the Pocket icon.

To see your saved articles, open the Pocket app. You can view articles that you've marked as favorites and that you've archived. You can also filter articles by keywords and tags. The article list can be shown as tiles or a list, and you can view all items, or just articles, videos, or images. You can also bulk edit items by tapping the Menu icon and selecting Bulk Edit. Tap to select all the items you wish to edit, then tap the icons at the top right corner to mark as read, mark as a favorite, delete, or tag all the selected items. Or you can cancel by tapping the Cancel button at the upper left.

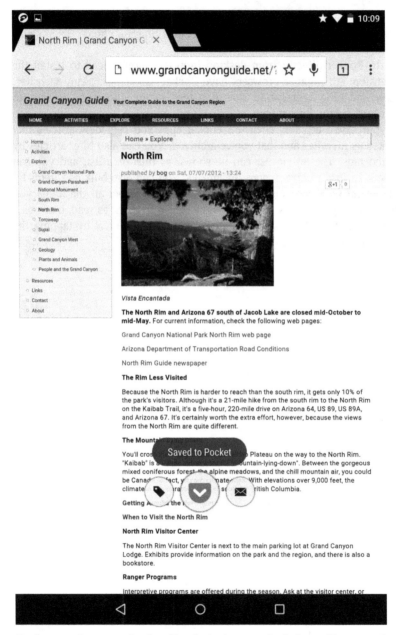

Saving a web page to Pocket. Touch the icon on the left to add tags to the article, and the icon on the right to email it

Making the Moon: The Practice Crater Fields of Flagstaff, Arizona

by **Jason Major**, universetoday.com
November 2

Apollo 15 astronauts David Scott and James Irwin practice LRV operations in

Reading a saved article in Pocket

Managing Files and Printing

There are many times you'll wish to move files around on your phone or tablet, or to move files to and from another mobile device or a computer. The easiest way to share files is via cloud storage, which is secure space on someone's server that you access via the Internet. Of course, to use a cloud, your device must be connected to the Internet via Wi-Fi or LTE.

If you don't feel comfortable sharing your files via a cloud because you're worried they might evaporate, you can transfer files directly between the Nexus and another mobile device or a computer via Wi-Fi.

Since the memory you have is never enough, you may find yourself running out of room, especially if you download a lot of videos and music. In that case, you'll need to remove unneeded files to make more space.

At some point you'll probably need to print a file from your device. You can do that two ways- by transferring the file to your computer and printing it, or printing directly from the device.

Google Drive

Google Drive (http://bit.ly/1dQZOo5) comes pre-installed on Android. Assuming you have a Google account, you have access to your files on any mobile device that can run the Drive app, as well as your computer via Drive for Windows or Mac. Drive is not available for Linux but you can access Drive from any web browser at www.google.com/drive. Drive gives you 15gb of free storage and you can buy more storage for a monthly fee.

Dropbox

Dropbox (http://bit.ly/1ho6XpN) is one of the oldest cloud services and probably still the most popular way to share files between computers and portable devices. After you install the Dropbox app, all you need to do is create a free Dropbox account. Then you can import or export files directly to Dropbox. Even easier, you can share files to Dropbox from most Android apps, such as camera apps, photo viewers, note taking apps, web browsers, social apps, and many more.

To access files on your phone or tablet, just install the Dropbox app on that device and log in to your Dropbox account.

You can install Dropbox for Linux, Mac, or Windows so that you can access Dropbox on your computer by downloading the appropriate version of the software from dropbox.com. This creates a special Dropbox folder on your computer. Any files you place in this folder are automatically added to Dropbox and can be accessed from any device running the Dropbox app. When you update a file, it is automatically synchronized.

You can also access your Dropbox files directly from dropbox.com in any web browser by logging into your account.

Running Android

Dropbox gives you 2gb of free storage, which is plenty for files being transferred between devices, or for temporary backup. If you need more space, you can buy storage from 100 to 500gb for a monthly fee.

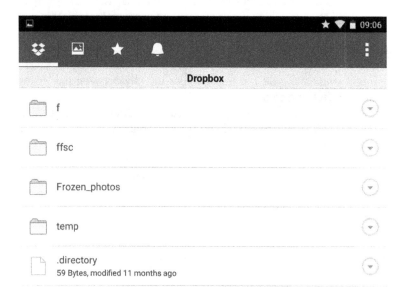

Files in Dropbox are accessible anywhere

Airdroid

Airdroid (http://bit.ly/193ML2o) is an Android app that lets you manage the files and folders on your Android devices from a web browser on your computer or anywhere you can run a web browser. After you install Airdroid, you'll need to create a free account and log in. Once Airdroid is running, it presents a web address that you type into the address bar in the web browser on your computer. An Accept button then pops up on the Android device. Tap this button to confirm the connection. You can now use the Airdroid web page to directly manage files on your device and upload and download files to your computer.

You can also manage apps, contacts, photos, music, videos, and use the cameras on the device from Airdroid on your computer. Some features, such as SMS messaging and call logs, only apply to Android phones.

Airdroid on a computer web browser

Wi-Fi File Explorer

Another wireless file transfer app, Wi-Fi File Explorer (http://bit.ly/HKYJZh) does only that, but does it very well. You can browse, transfer, download, upload, move, copy, and delete files on your Nexus using a clean drag and drop interface on your computer's web browser. You can explore all folders on your device, or display only pictures, music, videos, or camera photos.

ES File Explorer

ES File Explorer (http://bit.ly/1eqKKFq) is a free app that lets you manage files on your phone or tablet and share them with other people and other computers and devices. You can copy, cut, delete, and rename files. You can share files via Dropbox, Google Drive, most other popular cloud services, and FTP and Samba LAN. It also functions as an application manager and task killer.

Checking Storage Space

To see how much room apps and files are taking up on your device, swipe down from the upper right to open Quick Settings, tap Settings, then scroll to Device and tap Storage. After a few seconds, the Nexus calculates the storage taken by apps, pictures and videos, audio, downloads, cached data, and miscellaneous. To see details in each category, tap it.

The miscellaneous category can contain a lot of things. For example, if you download music from Amazon with the Amazon MP3 music app, the downloaded misc appears under miscellaneous. While it is possible to delete files from the detail screens, it's safer to delete files from the apps that use them.

Google Cloud Print

With Google Cloud Print (http://bit.ly/1d3rw4X), you can print directly from most Android apps to any printer attached to a computer that is on the Internet. Nearly all printers are supported when the computer is on and running Google Chrome. If you want to be able to print when your computer is not running, you'll need a Cloud Print ready printer. Currently, most Epson wireless printers and HP ePrint printers support Cloud Print.

Folder view in ES File Explorer

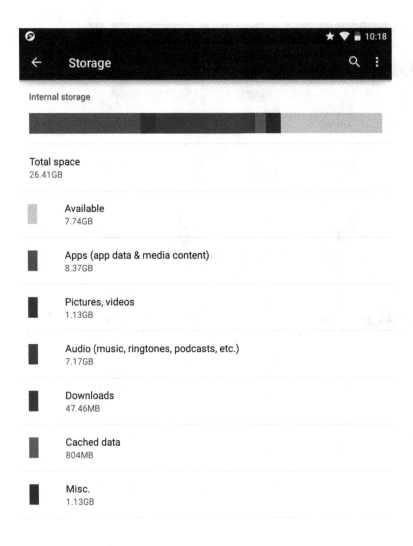

Storage used on a Nexus tablet. Tap any item to see more detail

Customizing Your Android

One of the great things about Android is that you can customize the user interface to work the way you want. The default Launcher that comes with Android and manages the Home screens is very flexible but you can gain even more control over the Home screens as well as the Apps Tray by installing a different launcher, such as Nova Launcher, which I'll discuss later in this chapter.

Of course, you have tens of thousands of Android apps to choose from, and that lets you replace nearly any of the stock apps with ones that have the features you need. Many apps come with their own widgets, and you can buy stand-alone widgets from the Google Play Store. Want a Home screen with a full page Google Calendar? Or do you want to turn your device into a beautiful nightstand alarm clock, so that you don't have to carry a separate travel alarm? Want a full page of weather? Or a main home screen with date, time, and weather? There are widgets to let you do all that and much, much more. I can only touch on the many ways you can customize your phone or tablet. The best thing to do is experiment. Nothing you do is irreversible.

That said, most Android phones and tablets have a user interface that is somewhat customized by the cellular provider or the tablet manufacturer, usually to simplify the interface or highlight provider features. Probably the most famous example is the Amazon line of Fire tablets. Amazon has customized the interface so much that they call it Fire OS. On Fire OS you can't install widgets, though you can place app icons on the Home screen and use app folders. At the other extreme, the Google Nexus line of tablets and phones run pure, stock Android as designed by Google.

Lock Screen

As I mentioned earlier, you almost certainly want to secure your phone or tablet by using a lock screen. See Securing Your Device for more information.

Widgets on the Lock Screen

Once you have a lock screen set up, you can customize it with widgets if your device is running Android 4 (Kitkat) or later. To customize your lock screen, drag right from the upper left side of the screen (this area is briefly shown when you wake up the device). Tap the large + icon to show the widgets available for the lock screen (not all are), then scroll up and down to see more widgets. Tap any widget to add it. For some widgets, you'll be taken to a configuration screen to set it up.

To remove a widget, long-press it and drag it to the Remove area at the top of the screen. When the widget turns red, lift your finger to delete the widget. You can add more than one widget to the lock screen, then swipe left and right to view them. The most useful widgets are ones that show date, time, weather, battery charge, and other status information that doesn't require interaction from the user.

Lock screen with custom wallpaper. Swipe up on the padlock to unlock the screen, swipe left on the camera icon to activate the stock Camera app, and swipe right on the micophone icon to start a voice search

Starting with Android 5 (Lollipop), widgets can no longer be added to the lock screen. But live wallpapers can display changing information- see below.

Home Screen

The Home screen can be customized in many different ways to suit your preferences. Some people like to see all the apps that they regularly use, while others prefer a cleaner screen. You can have more than one Home screen and access them by swiping, and you can install a new launcher to add more customization to the Home screens and Favorites trays.

Wallpaper

The background of the Home and lock screens can be changed by long-pressing a blank area of the screen until a dialog pops up. By default, you can choose from Photos, Wallpapers, and Live Wallpapers. If you have other wallpapers installed, you can choose from them also.

Static wallpapers consist of an unchanging image, while live wallpapers have animation and elements that update. An example of a live wallpaper is Weatherbug Elite (http://bit.ly/17cQxGQ).

Multiple Home Screens

The stock Android launcher has five home screens, which are accessed by swiping left and right. You can always return to the main Home screen by tapping the Home icon. You can add, remove, and rearrange apps and widgets on the Home screens. When you remove an app or widget from the Home screens, it is not removed from the device. You can always access all of your installed apps and widgets from the Apps Drawer.

Apps

Installing Apps

When you install an app from the Play Store, it always appears in the App Drawer, and may automatically appear on the Home screen, depending on your settings.

Adding Apps to the Home screens

Open the App Drawer and scroll to find the desired app. Then long-press and drag it to the desired home screen. The icon grid, which controls where icons will be, appears in the background. You can drag right and left to access other Home screens. Move the icon to the desired location and lift your finger. The icon will snap to the nearest opening on the icon grid. If another icon is in the way, bump it with the new icon and the other icon will move out of the way.

Moving Apps

Go to the Home screen with the app you wish to move, then long-press and drag it to the new position, bumping other icons out of the way if necessary. Then lift your finger.

Creating App Folders

You can organize apps into folders so they stack on top of each other and take up less room on the screen, both on the Home screens and in the Favorites Tray.

To create a folder, just move an app directly on top of another app until a blue circle appears around the icons. Then lift your finger. The folder is created, showing one app on top and the others behind. You can drag as many apps as you like into a folder.

To access the apps in the folder, tap to expand it. You can move the apps around within the folder by long-pressing them and dragging them. Remove an app from the folder by dragging it out of the folder to the Home screen. You can remove an entire folder when it is closed by long-pressing and dragging it to the X at the top of the screen.

You can also name the folder when it is expanded by tapping "Unnamed Folder" at the bottom of the folder, then entering a name with the virtual keyboard. This name shows up when the folder is closed. For example, you could move all your photography apps into a folder and name it "Photo".

Removing Apps

To remove an app from the Home screen, long-press it and drag it up to the X at the top of the screen until it turns red. Then release your finger. Again, this only removes the app from your Home screen and not from the device.

Uninstalling Apps

To remove an app from the device, open the Apps Drawer, find the app, and long-press and drag it to Uninstall at the top of the screen, then release the app. Tap OK on the confirmation screen.

Reinstalling Apps

All apps that you've installed from Google Play are archived indefinitely, and you can reinstall them at any time. This means that if you buy a new Android tablet or phone, you can reinstall any of your apps. Also, you only buy an app once and then can install it freely on as many Android devices as you have.

To reinstall an app, go to Google Play on the device or on your computer, find the app, and tap Install. Select the device if you have more than one. Apps that you own are marked with a check mark.

You can also install an app that you already own onto a different device by tapping Install. A drop down list shows all of your devices that are compatible with the app.

136

Widgets

Widgets are active icons of various sizes that can be installed on your Home screens and lock screen. Some apps come with widgets, while other widgets are bought separately from Google Play. Widgets can add a lot to the usefulness of your phone or tablet by showing active content such as clocks, the date and time, weather, battery charge, calendars, contacts, text messages, notes, social media apps, control buttons, music players, bookmarks, and much more.

Adding Widgets to Home Screens

Open the App Drawer and go to Widgets by tapping on Widgets at the top of the screen or swiping left until they appear. Long-press and drag the desired widget onto a Home screen and release it. Most widgets will present a configuration screen, which may be simple or complex depending on the widget. Some configuration screens have a Save icon which you must tap to save your configuration and create the widget.

Moving Widgets

Once a widget has been added to a home screen, you can long-press and drag it to a new position on the grid.

Removing Widgets

To remove a widget, long-press and drag it to the X at the top of the screen until it turns red. Then release it.

Some Widgets

Of course, there are thousands of widgets on Google Play. Here are just a few that I've found useful.

HD Widgets

HD Widgets (http://bit.ly/1iRTZMn) adds a set of gorgeous widgets of many sizes that can display the time, date, weather, and control switches. The color, background, transparency, and text are all customizable.

Android Pro Widgets

This app (http://bit.ly/HLilwa) adds widgets that can display bookmarks, calendars, your Facebook feed, a direct Facebook posting bar, contacts, and Twitter feed. The paid Pro version adds a combined Facebook and Twitter time line and many themes.

Beautiful Widgets

As the name implies, this set of widgets (http://bit.ly/195lzAj) brings a touch of beauty to your Home screens and it's one of the most customizable widget sets out there. Widgets from 1x1 to 5x5 are included for time, date, weather, control switches, and battery state. Each widget can be customized with hundreds of downloadable themes, and you can select the app that will run when you tap the widget. You can create a widget that is arranged just the way you want and controls your favorite apps.

Favorites Tray

The center of the Favorites Tray always contains the App Drawer, but apps can be added to either side for quick access. Just drag an app from the App Drawer to the Favorites Tray. To remove an app from the Favorites Tray, just drag it up to the X at the top of the screen until it turns red.

The three navigation buttons are always present at the bottom of the screen- Back, Home, and Last and can't be changed or customized.

Launchers

The user interface on Android is run by the default Launcher, which controls how you can place apps and widgets, how many Home screens you can have, and how the Favorites tray and Apps Drawer works. You can install many different launchers from the Play Store (http://bit.ly/1958Nln) to change the function and appearance of the user interface.

Nova Launcher

Nova Launcher (http://bit.ly/1aJ7A9G) adds many features to the stock launcher, including more ability to customize your Home screen, Favorites tray and Apps Drawer. With Nova Launcher, you can have as many Home screens as you want, accessed by swiping left and right. You can also have up to three Favorites Trays, also accessed by swiping the Favorites Trays left and right.

You can choose icon and color themes, set scroll effects, set infinite scroll so you can loop through all Home screens, change the style and background of folder icons, add multiple icons, hide apps, and change the size of the Home screen and Favorites Tray grids.

Nova Launcher Prime, the paid version of the app, unlocks even more features, including folders in the Favorites trays, add unread counts to email and text messaging icons, more scroll effects, and more. For the full list of features, go to Nova Launcher Prime in Google Play (http://bit.ly/1yspl6F).

Examples

To give you some ideas, here are several different ways to organize your device using Nova Launcher on a Nexus 7 2013 tablet.

Apps Galore

This is pretty much the default setup. The main Home screen has a time and date app that opens the Alarm app when the time is tapped, and the Calendar app when you tap the date.

Apps you use all the time, such as Gmail, Chrome, Contacts, and Google Maps are in the Favorites Tray.

Additional Home screens contain the apps you use most frequently, arranged in groups by function or alphabetically, or any way that works for you.

Widget Madness

This setup extends Apps Galore by adding large widgets to additional home screens, such as Facebook, Twitter, calendar, a note widget such as ColorNote or EverNote, Flipboard, and Feedly. By swiping the home screen, you can see recent activity in your social networks, keep an eye on upcoming events, see your notes, and keep an eye on the news without opening an app. Of course, tapping any of these widgets opens the full app.

Stacked Favorites

This is my current setup (see the Home screen shots below). At the top of the screen tehre are three customized Beautiful Widgets that display Date, time, and local weather. Tapping the date brings up the built-in Google Calendar app. Tapping the time starts My Alarm Clock, a nightstand digital clock that replaces a travel alarm. Tapping the weather starts NOAA Weather+.

At the side of the screen is a customized battery widget, part of the Beautiful Widgets set.

The rest of the main screen is occupied by named folders that contain all of the apps that I use most frequently. I'll add and remove apps from the folders as I get new ones or realize that I hard ever use an app.

The main Favorites Tray has the apps I use constantly- from left to right, a creen brigtness widget from HD Widgets, Smart Voice Recorder, DW Contacts, Evernote, Chrome, and Gmail. Not visible in the screenshot, the left Favorites Tray contains Merriam Webster Dictionary, Wikipedia, QickPic, The Photographer's Ephemeris, Backcountry navigator Pro, and Google Maps. The right Favorites Tray has Google News, Facebook, Twiteer, Feedly, Kindle, and Zinio.

The second Home screen has just three widgets for playing music- Amazon MP3, Google Play Music, and Jet Audio.

Taking Screenshots

Screenshots are obviously useful if you're writing a book about Android devices and need to illustrate your points! But they can be used for fun things too, like emailing a friend a picture of something you found on your phone. Fortunately Android makes it easy. Just hold down the power and volume down buttons simultaneously until the device makes a shutter sound. Screenshots are stored in the Screenshots folder under Pictures, and you can view and share them with others with the Photos app and other photo viewing apps. When viewing your screenshots, it can be easy to forget that you're looking at a screenshot, not the real screen, and then start to wonder why the device suddenly stopped responding to touch commands. If this happens, you can exit the screenshot by tapping the center of the screen to bring up the controls for the Photos app.

Home screen with custom wallpaper, date, time, and weather widgets, and app folders

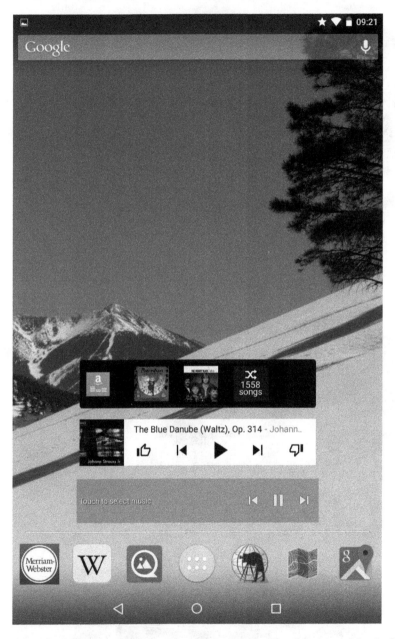

Second Home screen with Amazon MP3, Google Play Music, and Jet Audio music widgets

An open app folder

Finding Accessories

Most Android phones and tablets comes with just the device, a USB cable, an AC charger, and brief manual. You'll almost certainly want to protect your device with a case or cover, possibly a screen protector, add a vehicle charger, and maybe a physical keyboard and mouse. There are many such accessories and I can only give a few examples here.

Cases and Covers for Tablets

As an example, here are some cases and covers for the Nexus 7. Similar covers and cases are available for other tablets. If you prefer to use your tablet bare to keep the size and weight in your hand to a minimum, you can protect it in a case or sleeve when it's not in use. A simple sleeve such as the Leather Google Nexus 7 Sleeve by Dockem (http://amzn.to/1iS1u6a) may be all you need.

If you need more protection while using the Nexus, the ASUS New Nexus 7 FHD Official Travel Cover (http://amzn.to/1etoT01) protects the Nexus screen when closed, and wakes up the tablet when opened. Openings allow full access to ports, buttons, and the cameras.

You can also get covers which double as stands to make viewing movies and TV shows easier, such as the MoKo Nexus 7 FHD Case (http://amzn.to/1etqp2q).

Some cases can support the Nexus in either portrait or landscape orientation- an example is the rooCASE Nexus 7 FHD Dual-View Cover (http://amzn.to/1gwCDIa).

And finally, for really rugged protection, you can't beat Otterbox (http://amzn.to/1gwCS5T). This case protects the ports as well as the screen.

Phone Cases

Most users protect their bare phone with a case. An example is the Incipio MT-298 DualPRO (http://bit.ly/1CDnhvH). This slim case doesn't add much bulk so you can still carry your phone in a pocket. As with tablets, Otterbox is widely regarded as making the toughest and most protective phone cases. For the Droid Maxx, the Defender series (http://bit.ly/1yWbCoW) offers the most protection. As with tablet cases, the Otterbox includes screen protection. If you carry your phone on your belt or if you work outdoors, an Otterbox is a good choice. The Otterbox Commuter series (http://bit.ly/1zGOXQi) are less bulky and don't include a belt clip- but still have screen protection.

Screen Protectors

If your tablet or phone case doesn't include a screen protector, you can buy one separately. For the Nexus 7 and my phones, I've used the protectors from JETech (http://bit.ly/1z4tLUJ). While most phones and tablets use very tough screens, they can still get scratched, especially if you carry your device in a purse or pocket. Tests have also shown that screen protectors add some protection against broken screens if the phone is dropped.

To avoid annoying air bubbles and bumps in the screen protector, follow the manufacturer's directions exactly when installing a protector. The process is tedious but an installed protector lasts a long time.

Headphones

While many Android phones and tablets have decent speakers for such a small device, you'll probably want headphones if you listen to a lot of music, play games, watch movies and TV shows, or use Skype. Good quality headphones also produce better sound than small speakers. Of course, you should use headphones in places such as libraries, airliners, and other public places to avoid disturbing others.

There are three basic types of headsets- in the ear "earbud" style, on the ear, and over the ear. Any of these types can be wired or wireless using Bluetooth. You can also buy headsets with active noise cancellation.

Wired Earbuds

Wired earbuds are the lightest and most compact headphones, and good ones produce surprisingly good sound. They also use the least power. Low end earbuds such as JLab Jbuds (http://amzn.to/19e7pge) are just a few dollars. High end earbuds such as the Sennheiser CX 300B MK II (http://amzn.to/1cfSmG0) are more expensive but produce better sound.

On the Ear Headphones

These headphones are larger than earbuds and sit on your ear rather than inside. Still light and compact, these are a good choice for kids because it's harder to play music at ear damaging levels. Some models have volume limiters as a further level of protection. Another option is a built-in microphone, which is very useful if you make audio and video calls with Skype or want to use the headset with your phone. An example of an on the ear headset with mic is the wired JLab INTRO Premium On-Ear Headphones (http://amzn.to/1800USc).

Over the Ear Headphones

These produce the best sound and also reduce external noise more than earbuds and on the ear headphones. Although they are heavier and bulkier than the others, they're worth it in noisy places such as airliners, trains, and cars (assuming you're not driving the plane, train, or automobile!). An example is the wired Sennheiser HD 419 Headphones (http://amzn.to/1bkOEcy).

Active Noise Canceling Headphones

You can also get headphones with active noise cancellation. These headsets use tiny microphones to pick up ambient noise, and then generate an out-of-phase sound wave to cancel the external noise. Active cancellation is most effect at lower frequencies such as that produced by engines and wind noise. If you fly and listen a lot, they are pretty hard to beat. The pioneer in active noise canceling headphones was Bose, and their aviation headsets for pilots are still the gold standard. Fortunately Bose's music headsets, such as the Bose QuietComfort 25 Headphones (http://bit.ly/15s9U3T) are a lot less expensive than the pilot headsets, and they work stunningly well.

Bluetooth Headphones

These headphones use a Bluetooth wireless connection to your phone or tablet. The advantage is that you have no wires and plugs to contend with. The disadvantages are higher cost, limited range, and increased battery drain, both on the Nexus and the batteries that power the headphones. Bluetooth has a range of about 20 feet, so you'll need to carry the tablet with you if you're moving around or stay near it to avoid audio dropouts.

Some Bluetooth headphones also have a mic for use with Skype, Google Voice, voice recorders, and other apps that use voice input- and you can also use them to make calls with your Android phone.

In the Ear Bluetooth Earbuds

An example is the Creative WP-250 Active Bluetooth Headphones with Invisible Mic (http://amzn.to/1cfZbaC).

On the Ear Bluetooth Headphones

An example of a headset with microphone is the MEElectronics Air Fi Runaway Bluetooth Stereo Wireless Headphones (http://amzn.to/16WnFps).

Over the Ear Bluetooth Headphones

The Creative WP-350 Wireless Bluetooth Headphones (http://amzn.to/1aAKpbP) also features a mic for use with your phone or tablet.

Speakers

As with headsets, you can get wired or wireless Bluetooth speakers for use with your phone or tablet. There's not much advantage in wired speakers over the built-in speakers, unless the speakers are powered.

The Creative D100 Wireless Bluetooth Speaker (http://amzn.to/1iZTErd) works with any Bluetooth device including Android phones and tablets as well as other Bluetooth audio devices. This speaker also uses NFC technology, so you can pair it with many Android devices by tapping the back of the device on the speaker. It also has a standard 3.5mm headphone jack to connect to audio sources via a cable.

Chargers

Although most Android phones and tablets come with a USB cable and AC adapter for charging, you may want a second charger for travel (http://amzn.to/17ISsJm). If you travel a lot, you may need a car charger. Multiport chargers are best since you can charge more than one devices, such as your phone, at a time. I've used the iFlash High Output 4 USB Car Charger (http://amzn.to/19L5zYK) with good results.

For extended wilderness travels, such as river trips and backpack trips, you should consider a solar charger, such as the Suntactics sCharger-5 (http://amzn.to/1cOk8OG).

Vehicle Mounts

The best vehicle electronic mounts without question are made by RAM. They aren't cheap but they'll keep your device safe. A mount that works with 7-inch tablets is the RAM Mount X-Grip Universal Tablet Holder (http://amzn.to/HLsSro). Other sizes are available for larger tablets, as well as phones. You'll also need a mount such as the RAM 166U Double Ball Twist-Lock Suction Cup Mount (http://amzn.to/16N5P8d) to attach it to your vehicle. There are many other mounting options, including screw-down, permanent mounts. These keep your device secure even on rough dirt roads.

Keyboards and Mice

To make text-intensive work easier, you can connect keyboards and mice to your phone or tablet. Though its possible to connect USB keyboards and mice using cables, a hub, and an adapter, Bluetooth wireless devices are much more convenient.

Wired USB keyboards and mice require a USB to microUSB adapter as well as a USB hub in order to connect both devices to the single USB port on the Nexus. You could use a wireless USB keyboard and mouse that use the same USB receiver, which avoids the need for a hub. But you still need the adapter to connect the USB receiver to the device. In addition, using USB mice and keyboards drains your battery significantly faster, so you may need to keep the Nexus on a charger. You can't do that while using USB devices. For these reasons I recommend Bluetooth keyboards and mice.

An example of a mouse: the Gear Head Blue Tooth Laser Mouse (http://amzn.to/17f7plE), which works well on my laptop, Nexus 7, and Droid

Maxx. And the Azio KB334B Mini Wireless Bluetooth Keyboard (http://amzn.to/HCtQpl) has also worked well for me.

If you use a keyboard a lot, you might want to consider a case with a built-in keyboard such as the Bluetooth Keyboard Stand Case (http://amzn.to/HDcjNx) for the Nexus 7. With the addition of a Bluetooth mouse, this type of case turns the Nexus 7 into a very light and compact notebook computer.

The Bluetooth radio in the phone or tablet goes to sleep to save power when Bluetooth devices are inactive. Although Bluetooth wakes up as soon as you start typing or move the mouse, there may be a slight lag until Bluetooth wakes up. You can tell that Bluetooth is awake because the gray Bluetooth icon in the Status area at upper right will turn blue.

Extended Warranties

If your phone or tablet is exposed to situations where it could be lost or damaged easily, you might want to consider an extended warranty plan. As an example for the Nexus 7, SquareTrade offers 1 to 3 year plans (http://amzn.to/1gwFQYk). These plans protect the Nexus from drops, spills, and hardware failures, and guarantee your Nexus will be repaired or the full replacement cost received within 5 days. You get free two-way shipping for repairs.

Maximizing Battery Life

White many Android phones and tablets have respectable battery life right out of the box, there are steps you can take to extend it. You'll want to do this if you won't be able to charge the device for a long period, such as a long flight or in a backcountry situation.

Checking Battery Usage

Swipe down from the top of the screen with two fingers to show Quick Settings. Then tap the Battery icon to show battery drain arranged in order from highest to lowest use. Tap on any item to show details and suggestions on reducing battery drain. You don't have control over some apps, such as Android System and Android OS, that are always running. But you do have some control over others, such as the screen, Wi-Fi, and Bluetooth.

Battery Saving Apps

There are a lot of battery-saving apps on the market that claim to extend battery life, but my experience has been pretty dismal, both on my Nexus 7 and my Android phones. The apps I've tried either do nothing, or in some cases actually drain the battery faster than before.

An app that finally shows some promise is DU Battery Saver (http://bit.ly/1EfOjLo). It has an Optimize button that automatically shuts down idle apps, and a battery monitor that lets you shut down apps manually. There is also an Optimize widget that you can use to shut down apps without opening DU Battery Saver itself. I suggest you test the free version extensively before buying the paid Pro version.

Display

The largest power user that you have control over is the display. There are two settings that affect how much power the display uses, brightness and sleep time.

The brighter the display is, the more power it uses. It's a good idea to keep Adaptive Brightness, the default, turned on, so the device can adjust the brightness to the ambient light. If you manually set the display to the brightest setting so you can read it in bright light, it uses a lot more power.

The other setting, sleep timeout, sets the amount of time the tablet is idle before the screen blanks. The default of 1 minute is good in most situations, but if your battery is low or you won't be able to charge your device for a while, try setting it to 15 or 30 seconds. On the other hand, if the phone or tablet is plugged into a charger, you can set the timeout for up to 30 minutes so you're not having to constantly use the power button to wake it up.

To change brightness settings, open Quick Settings, then drag the Brightness slider to manually set brightness. To turn Adaptive Brightness off or on, open Settings, then Display. You can also change display timeout from this screen.

Animated Wallpaper

Animated wallpaper looks cool but uses processor power. Turn it off by opening Settings, then Display. Tap Wallpaper and choose any wallpaper other than Active Wallpaper.

Airplane Mode

The radios that your phone or tablet use to connect to Wi-Fi and LTE use a lot of power, especially in areas where the signals are weak. The Wi-Fi transmitter keeps trying to maintain a connection, and LTE transmitters increase power in fringe areas in an attempt to stay connected.

Bluetooth also uses a radio to connect to Bluetooth devices such as headphones, speakers, mice, and keyboards. Bluetooth is actually pretty efficient, because the radio goes to sleep after a few seconds of no activity from Bluetooth devices. Still, keep Bluetooth off unless you are using a Bluetooth device.

The status of the Bluetooth radio is shown in the Status area at the upper right corner of the screen. A blue Bluetooth icon means that Bluetooth is on and active. A gray icon means that Bluetooth is sleeping and that it will wake up when you use a connected device. No icon means Bluetooth is off or the Nexus is in Airplane mode.

You can greatly extend battery life, especially while traveling, by switching your device to Airplane mode. This shuts down all of the radios in the tablet, including Bluetooth. To switch to Airplane mode, pull down Quick Settings, then tap the Airplane Mode icon. Do the same to turn Airplane Mode off.

To turn off Bluetooth only, pull down Quick Settings and tap the Bluetooth icon.

GPS

Android devices use up to three different services to determine your location, the Wi-Fi network, the cellular network if you have an LTE tablet or a phone, and GPS. Even though GPS uses a receiver only, it uses the most power of the location services, so turn GPS off if you don't need precise location.

The reason GPS uses so much power is that the system uses very weak radio signals from satellites orbiting 12,000 miles above the Earth to determine your position. The GPS receiver in the Nexus is a sensitive receiver that uses a lot of power, and calculating your position uses a lot of processor power.

The GPS icon appears in the Notifications area at the upper right corner of the screen when GPS is in use. While most apps use GPS only when they really need it, some apps leave GPS on even when the app is running in the background. If the GPS icon stays on for long periods of time, you may want to disable GPS to extend battery life. Or you can change the GPS settings in the app itself.

To turn off GPS, open Settings, then tap Location. You can turn all location services off or on, or you can set Location Mode to High accuracy (uses GPS, Wi-Fi, and LTE), Battery saving (uses Wi-Fi and LTE only), or Device only (uses GPS only).

Google Maps is an example of an app that turns off GPS except when finding your location or navigating to a location. Backcountry Navigator Pro, on the other hand, tends to keep the GPS on even when the app is in the background. If you're just viewing maps and not actually navigating with GPS, you can turn GPS off in Backcountry Navigator. Tap the Menu icon at the upper right corner, then Settings. Then tap Keep GPS On to toggle this setting.

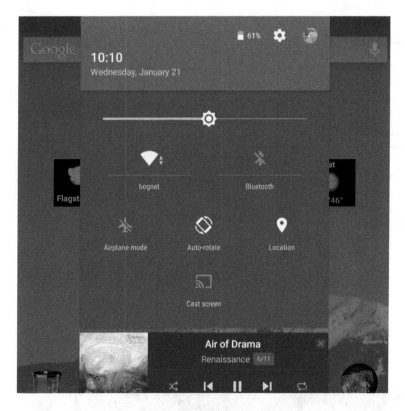

Quick Settings, activated by swiping down from the top of the screen with two fingers, gives access to commonly-used settings. The available settings vary between devices

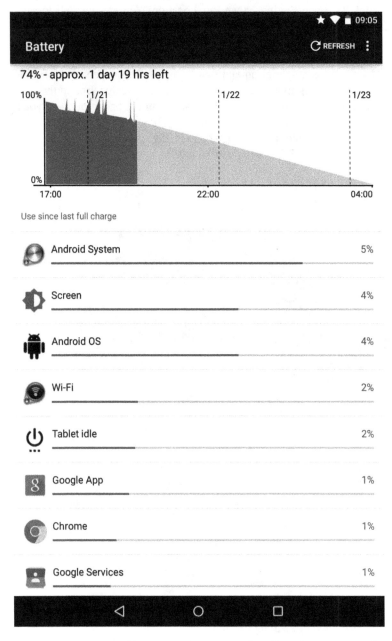

Battery status screen. Tap on any item to see details and suggestions for reducing power use

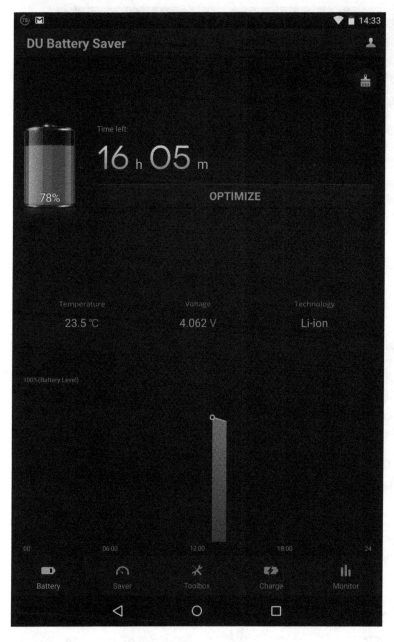

DU Battery Saver

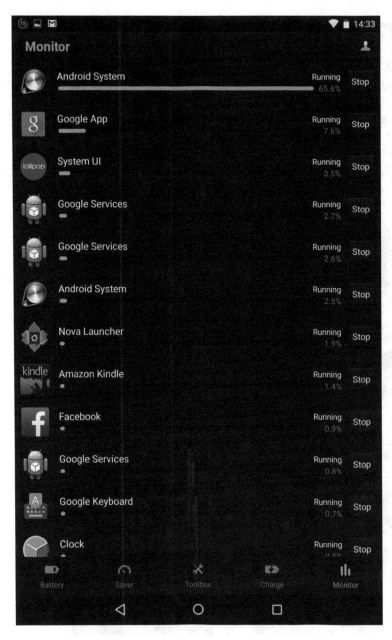

The Monitor page in DU Battery Saver lets you see how much power each app is using and manually shut them down

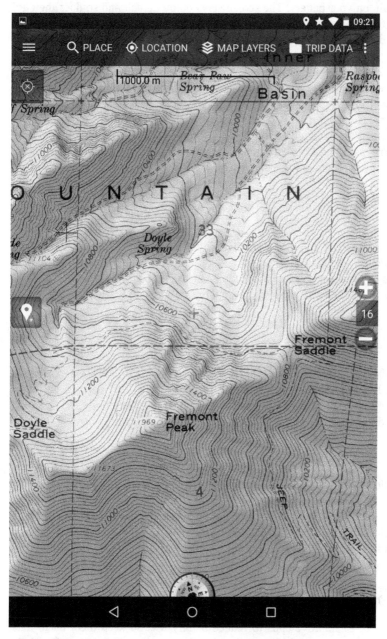

GPS is active, as shown by the location icon left of the Wi-Fi icon on the status bar at the top right corner of the screen

Navigation

Using your Android device to navigate by street or in the backcountry is by far the largest drain on your battery. The screen stays on by default and so does the GPS receiver. If you plan to use your device for extensive street navigation in a vehicle, you should power the tablet from the vehicle with a car charger. See the section Chargers.

In the backcountry, you can greatly extend battery life by leaving your phone or tablet off except when you need to check your location and progress. By doing this you can extend the battery life to days rather than hours. In the case of a phone this also preserves your battery for emergency calls. Don't forget to download maps to your phone or tablet at home so you'll have access to them when you're out of Wi-Fi or LTE range. See the section Topographic Maps for an example. Another option for extended backcountry trips is a solar charger- see Chargers.

Video

Watching movies and TV shows is power-intensive because the display has to be on continuously and streaming video uses a lot of data, which makes the Wi-Fi or LTE radio in the device work harder. You can extend battery life while watching videos by keeping the display set to adaptive brightness as described above in Display.

Audio

Audio only uses a lot of power if you use your device to drive external wired unpowered speakers or leave the display on, but there's little need to do either. Powered speakers and Bluetooth speakers have their own batteries or AC adapters and don't drain devices much faster than wired earbuds.

Remove Unneeded Apps and Widgets

You should always remove apps that you no longer use. They not only take up storage space on your phone or tablet, they use processor and battery power, especially if left running in the background. To remove an app, find it in the Apps Drawer, then long-press it and drag to Uninstall at the top of the screen. Home screen widgets use some power to update themselves, so don't place widgets that you don't use.

Keep Apps Up to Date

Developers often improve apps to reduce battery use, so keep them up to date. Some apps update automatically, while others require your permission. Check the

Notification area at the upper left corner of the screen for apps that are waiting for permission. If you have an LTE tablet or a phone, you can also let apps update only via Wi-Fi, which uses less power than LTE (and avoids using your data allowance.)

Reduce App Polling

Some apps, such as Facebook, periodically check for updates. Not all apps allow you to set the interval but for those that do, increasing the update interval uses less battery power. Some news and RSS feed apps also let you change the update frequency.

Turn off Flash Video

Flash video uses a lot of power, which is apparently why Adobe has discontinued Flash support for mobile devices. So don't install Flash unless you need it- and if you do use it, set your Flash-compatible browser to only load Flash when it needs it. For example, in Puffin browser (http://bit.ly/1cy3mTZ), tap the Menu icon at the upper left corner, scroll down and tap Settings, then check that Load Flash is set to Auto.

Doze Mode

Doze Mode was added in Android Marshmallow. It works by sensing when your phone is not in motion and hasn't been used for a while, and then putting the phone into deep sleep, almost Airplane mode. Doze Mode is automatic and not something you can control.

Resetting your Device

If battery drain still seems excessive, you can try resetting the device. This shuts down unnecessary apps. See Resetting Your Device.

Solving Problems

Back it Up!

As with any computer, the best defense against problems is a good backup. If your device fails or you have to reset it to factory defaults, having a backup means you can pick up right where you left off. Fortunately, as of Marshmallow, Android backs up everything, including app data and settings (the app developer must enable this feature, so not all apps will be backed up.) To check that backups are on, open Settings, then Backup and Reset. Check that Back Up My Data is On.

Resetting your Device

Many problems, such as apps freezing or operating erratically, can be resolved by a soft reset. In fact, you should periodically do a soft reset on all your Android devices, even if you aren't having a specific problem. A soft reset clears memory and closes down idle apps, and a soft reset does not affect stored data or installed apps.

To do a soft reset, hold down the power button until the power off dialog appears, then tap it to shut down the device. After the screen goes blank, press and hold the power button until "Google" appears on the screen. After a minute or so the Home screen should appear.

Checking Memory Use

Apps that hang the system or run very slowly may be using excessive memory. Starting with Android Marshmallow, you have a way to check memory use. Go to Settings, About Phone, Memory. You can select how far back to look at memory use. To stop an app, tap the app, then tap the "i" icon, and tap "Force Stop."

Wrong App Opens

When more than one app can handle a request from another app, such as sharing a photo from a photo album, you are asked which app you want to use, and whether you want to use this app always, or just once. If you tap "always", either on purpose or accidentally, that app will be used by default and you won't be asked again. If you want to use a different app, you'll have to clear the default.

To do so, go to Settings, Device, Apps, and tap the app that is being opened by default. Then tap the Clear Defaults button. Next time you'll again be asked which app you want to use for the action. Select the app and tap "Always" if you want to always use that app, or "Just once" if you want to be asked each time.

Device Won't Charge or Turn On

First, attempt to charge the device with the charger that came with it, or a charger rated at least 1.35 amps. Don't use a computer USB port unless it's marked as a charging USB port- standard USB ports supply only 0.5 to 0.9 amps. A charger rated less than 1.35 amps will charge phones and tablets very slowly, or not at all, especially if the display is on.

If the device is off when you plug in the charger, a battery icon should appear on the screen within a minute showing that it is charging. If the device is on, the battery icon in the Status area at the upper right corner should change to show a lightning bolt, indicating that it is charging.

If the device isn't charging, try another power outlet, charger adapter, and cable. If the battery icon appears, press and hold the power button at least 15 seconds to see if the device powers on.

If the charging icon doesn't appear, leave the device charging for at least an hour, and then try powering it on by holding the power button at least 15 seconds.

And finally, if your still doesn't power on, make absolutely certain your charger is functioning by plugging in another device that uses a micro USB port for charging, such as a phone, and making sure that device is charging. Then reconnect the problem device to the charger, leave it overnight, then try powering it up by holding the power button down for at least 15 seconds.

If this fails and you have a Nexus device, check for updated troubleshooting steps at Google Play Help: http://bit.ly/1znbMs2. US customers can also call Google Hardware Support at 855-836-3987, 24 hours per day, 7 days a week. See http://bit.ly/180xt2o for phone numbers and hours for other countries. If you have a tablet bought from a retailer (not a cellular carrier), contact the manufacturer. If you have a phone or a tablet that you bought from a cellular carrier, contact the carrier's technical support.

Product Forums

Another place to look for help on a specific problem with a Nexus device is the Google Nexus Product Forums (http://bit.ly/1dZvkoX). This is also a good place to look for general help with using the Nexus.

Most other Android phone and tablet manufacturers have product forums. Just use a search engine to find them.

Factory Reset

This is a "hard" reset that removes all user data and apps, and restores the Nexus to the factory defaults. You should never have to do this to recover from a frozen app or unresponsive screen. Mostly you want to do a factory reset if you sell or give your Nexus to someone else.

To do a factory reset, swipe down from the top of the screen with two fingers to open Quick Settings, tap Settings, and scroll down to Backup & Reset. Tap

Factory data reset and follow the prompts. The Nexus will power down and restart. You'll be taken to the welcome screen and guided through the setup process.

Also by the Author

Books

The Complete 2015 User's Guide to the Amazing Amazon Kindle Fire (with Stephen Windwalker)

The Complete 2015 User's Guide to the Amazing Amazon Kindle - E Ink Edition (with Stephen Windwalker)

Publish! How To Publish Your Book as an E-Book on the Amazon Kindle and in Print with CreateSpace

Exploring With GPS: A Practical Field Guide for Satellite Navigation

Grand Canyon Tips: The Local's Guide to Avoioding the Crowds and Getting the Most Out of Your Visit

Grand Canyon Guide: Your Complete Guide to the Grand Canyon

Exploring Great Basin National Park: Including Mount Moriah Wilderness

Google Nexus 2013: Making Your Android Tablet Work for You

Websites

BruceGrubbs.com

BrightAngelPress.com

GrandCanyonGuide.net

ExploringGreatBasin.net

ExploringGps.com

FlagstaffFigureSkatingClub.com

Blogs

Get Out and Stay Out

Travels With Kindle

Social

linkedin.com/in/brucegrubbs

facebook.com/bruce.grubbs.outdoor.author

twitter.com/grandcynwriter

Before You Go

Please consider leaving a review for this book on Amazon at amzn.to/1mqLTCV. I would greatly appreciate it.

And, if you want to be the first to know about my new books, as well as revisions, consider signing up for my mailing list at eepurl.com/bPW7dD. This list will only be used for that purpose- I will NEVER spam you or share the list with anyone. You can unsubscribe at any time.

About the Author

The author has a serious problem- he doesn't know what he wants to do when he grows up. Meanwhile, he's done such things as wildland fire fighting, running a mountain shop, flying airplanes, shooting photos, and writing books. He's a backcountry skier, climber, figure skater, mountain biker, amateur radio operator, river runner, and sea kayaker- but the thing that really floats his boat is hiking and backpacking. No matter what else he tries, the author always comes back to hiking- especially long, rough, cross-country trips in places like the Grand Canyon. Some people never learn. But what little he has learned, he's willing to share with you- via his books, of course, but also via his websites, blogs, and whatever works.

Index